D0372938

Henry M. Morris, Ph.D.

CREATION-LIFE PUBLISHERS
San Diego, California 92115

THE REMARKABLE BIRTH OF PLANET EARTH

This English edition published in 1978 by Bethany Fellowship, Inc., by special arrangement with Creation-Life Publishers.

ISBN No. 0-87123-485-8

Printed in the United States of America

TABLE OF CONTENTS

PREFACE

This book has been written in order to give busy, but interested, readers a brief summary of both Biblical and scientific reasons for believing in creation instead of evolution. The origin and early history of the earth and man is a marvelous and fascinating story of God's great power and foresight, given by revelation in the Bible and now strikingly confirmed by modern science.

The theory of evolution has dominated our society, especially the schools, for almost a hundred years, and its influence is largely responsible for our present-day social, political, and moral problems. Many people today, including scientists, are again examining the evolution-creation question and often are amazed to find that evolution is merely an unreasonable theory containing many scientific fallacies. Creation, on the other hand, is a scientific theory which *does* fit all the *facts* of true science, as well as God's revelation in the Holy Scriptures.

There are hundreds, perhaps thousands, of scientists today who once were evolutionists but have become creationists in recent years. I myself was one of these, having accepted the evolutionary theory all through college. Since that time, however, as a result of considerable reading in all the various sciences which bear on the evolution-creation question, as well as in the Bible, I personally have become thoroughly convinced that the Biblical record, accepted in its natural and literal sense, gives the only scientific and satisfying account of the origin of all things. Many other scientists today can give a similar testimony.

Such a conclusion obviously has profound personal meaning as well. If God created all things, then He also controls all things, and has a purpose for all things—including you and me! History is not a blind interplay of forces, but is the outworking of the Creator's plan. Things that seem to be wrong with the world—war, disease, pollution, hatred, poverty—are not permanent residents, but only temporary intruders, and will someday be exiled forever.

The record of creation has thus become also the prophecy of restoration, for God cannot fail in His creative purpose! Evil has been allowed for a time, since otherwise man (created in God's image) could never have experienced either responsibility or reconciliation.

It is time, therefore, for "all men everywhere to repent" (Acts 17:30). Repentance means essentially a change of mental attitude, rejecting the man-centered philosophy of struggle and evolution and accepting instead the God-centered truth of creation and redemption.

The purpose of this book is to give a solid, though brief, treatment of the major aspects of the question of origins. The supposed evidences for evolution are shown to be capable of more satisfactory interpretation in terms of creation, and the unresolved difficulties of the evolutionary theory are shown to be positive evidences for creation.

The discussion is primarily approached from the Biblical point of view, and assumes throughout that the Bible is the Word of God, divinely inspired and, therefore, completely reliable and authoritative on every subject with which it deals. To some readers, of course, this may seem an unwarranted assumption. I can only urge them to approach the discussion with an open mind, learning as a matter of interest just what the Bible does have to say on these subjects.

I think they will be pleasantly surprised to find how relevant and accurate the Bible actually is when dealing with such matters. The discussions in the book also include enough scientific information to verify the historical and scientific confidence we are placing in the Scriptures. For more detailed scientific documentation the reader is referred to the recommended reading list in Appendix C.

In closing these opening remarks, I want to list seventeen summary statements which, if true, provide abundant reason why the reader should reject evolution and accept special creation as his basic world-view. The evidence that they *are* true will be found in the pages that follow.

1. The Bible clearly teaches that all things were created in six natural days several thousand years ago, and all other Biblical interpretations of the creation account contain many irreconcilable contradictions with both science and Scripture.
2. There is no demonstrated *fact* of science which cannot be satisfactorily correlated with this simple and straightforward Biblical record.
3. There are many amazing scientific insights in the Bible which long preceded their actual discovery by modern scientists.
4. No fact of actual observation has ever confirmed the *general* theory of evolution, as distinct from those minor variations which are known as the *special* theory of evolution.
5. All facts of observation support the concept of the original creation of distinct kinds of organisms, each with a genetic ability to develop into many different varieties but never into a new kind.
6. Both the present world of living organisms and the fossil world of dead organisms exhibit the

same clearcut gaps between kinds, with no transitional forms between.

7. Man has never created "life in a test-tube", nor is there any evidence that non-living substances have ever evolved into living organisms, either in the past or in the present.
8. Inheritable and novel changes (mutations) which take place in organisms today have always been observed to be harmful in some way to the organism in its natural environment.
9. The tremendous complexity and order of the world and its plants and animals can only be explained by intelligent planning, not by a random process of chance variation and natural selection.
10. The basic laws of nature, to which all natural processes must conform, are laws of conservation and disintegration, as the Bible teaches—not laws of innovation and integration, as evolution teaches.
11. The entire fossil record can be explained better in terms of cataclysmic destruction of all the ecological zones of one age than in terms of evolutionary development of changing world-wide floras and faunas through many ages.
12. There are more natural phenomena indicating the earth is very young than those indicating it is old, and all the latter can easily be reinterpreted in terms of young age.
13. Belief in special creation has a salutary influence on mankind, since it encourages responsible obedience to the Creator and considerate recognition of those who were created by Him.
14. Belief in evolution is a necessary component of atheism, pantheism, and all other systems that reject the sovereign authority of an omnipotent personal God.

15. Belief in evolution has historically been used by their leaders to justify a long succession of evil systems–including fascism, communism, anarchism, nazism, occultism, and many others.
16. Belief in evolution and animal kinship leads normally to selfishness, aggressiveness, and fighting between groups, as well as animalistic attitudes and behaviour by individuals.
17. Belief in evolution leads usually and logically to rejection of the trustworthiness of the Bible and, therefore, failure to appropriate its promises leading to salvation and eternal life.

The reader will find ample evidence for the foregoing statements in the eight chapters of this book, and much more evidence in the books recommended for further study. The evidence clearly shows creation to be a much more reasonable alternative than evolution. That being the case, the reader should soberly consider the consequences in light of eternity should he choose to continue to reject (or, which is even worse, *neglect)* his Creator and Redeemer in deference to the unproved philosophy of evolution.

"Know ye that the Lord He is God: it is He that hath made us, and not we ourselves: . . . be thankful unto Him and bless His name" (Psalm 100:3,4).

Henry M. Morris
San Diego, California
September 1972

CHAPTER I
THE AMAZING ORDER OF COMPLEX THINGS

One of the strangest phenomena of human nature is that brilliant minds often make foolish decisions. There is no more common and universal fact of experience than the fact that order never arises spontaneously out of disorder and a design always requires a designer. Yet many scientists and other intellectuals believe that our intricately-designed and infinitely-ordered universe developed all by itself out of primeval chaos!

The law of cause-and-effect is a universally-accepted concept of modern science. If like causes did not produce like effects, in fact, science would be impossible. The world would be a chaos, not a cosmos.

Furthermore, since an infinite chain of secondary causes extending chronologically backwards into past eternity is a meaningless and unsatisfying alternative, the concept of a great uncaused First Cause is the most reasonable and powerful explanation for the infinite array of ordered effects which are observed in the universe.

Thus the simplest, and yet most profound, explanation for the world is found in the first verse of God's Word: "In the beginning, God created the heaven and the earth." The tremendous complex of orderly relationships in the universe, challenging the highest intelligence of man to describe, certainly implies an intelligent First Cause of these relationships. It seems absurd in the very nature of things for men to use their own intellectual capacities to try to devise an intelligent explanation of things in terms of non-intelligent origin. How could unthinking atoms create intelligent thought? No wonder the Biblical writers are emphatic in denouncing such sophistries. "The fool hath said in his heart, There is no God" (Psalm 14:1). "Professing themselves to be wise, they became fools, . . . and worshipped and served the creature more than the Creator, . . . even as they did not like to retain God in their knowledge" (Romans 1:22,25, 28).

If the law of cause-and-effect applies to all known phenomena—and it does—then there must be an adequate cause even to explain why such men seek to explain all things without a First Cause. Atheists are not born that way; it is natural and normal for a child to believe in God, and this innate belief must be "educated" out of him if he is to become an atheist.

The cause of this phenomenon is to be found in the words of Romans 1:28, above quoted: "They did not like to retain God in their knowledge." It is not that there is not adequate evidence of a Creator, it is just that men prefer to avoid Him. They *make* themselves believe in "no God" instead of retaining their inborn belief in His true existence.

Therefore they try by every means possible to explain away the innumerable evidences of order and design in nature on the basis of some natural process, without creation. But how can complexity arise naturally out of simplicity, and order out of disorder? Nothing like this is *observed* to happen in the real world. *Real* processes always go, if left to do what comes naturally, in the direction of greater disorder and randomness. How can the *appearance* of design be produced without the *reality* of design?

This argument from design has always been considered as a most powerful evidence for the existence of God. A watch requires a watchmaker: then, what about the far more intricate and precise atomic and sidereal clocks? A water supply system requires the efforts of a great many skilled engineers and builders; then, what about the marvelous reservoir, pumping, purification, and distribution system involved in the hydrologic cycle which supplies water to the earth's inhabitants? A great building presupposes a trained architect; the infinitely-more complex structure of the human body cannot even be analyzed, let alone designed, by man. The greatest digital computers are absurdly simple in comparison to the complex circuitry of the human brain and nervous system.

The idea that a complex structure or system can somehow be formed by chance is a persistent delusion accepted by evolutionists. Typical naturalistic reasoning supposes that anything can happen if enough time is available. Monkeys pecking away blindly at typewriter keys are bound eventually to hit on a Shakespearean sonnet, so the thinking goes.

But this idea is absurd. To illustrate, consider an ordered structure of, say, 200 parts. This is not an unusual number—the human skeleton, for example, contains more than 200 separate bones, all aligned together into a perfectly integrated functioning whole. There are innumerable systems in the world far more complicated than this.

Consider the possible number of different ways 200 parts could be aligned together. A system of one part could be lined up in only one way; one of two parts in two ways (1x2); one of three parts in six ways (1x2x3); one of four parts in 24 ways (1x2x3x4); and so on. Thus a system of 200 parts could be aligned in a number of different ways equal to 1x2x3x4x5x6 . . . x . . . x 200. This number is called "200 factorial" and is written "200!".

This is a tremendously large number. It can be shown to be approximately 10^{375}, that is, a number written as "one" followed by 375 "zeros". Therefore, the *correct* alignment of the 200 parts has only one chance out of 10^{375} of being selected on the first trial.

Suppose a new trial can be made every second. In all of supposed astronomic time (about 10 billion years) there have only been 10^{18} seconds, so the chance that the correct alignment might be obtained once in the 10 billion years would be only one out of $10^{(375-18)}$, or 1 in 10^{357}. This is still practically zero.

Suppose that we try to improve the chances by arranging to have a large number of sets of the 200 parts, all being tried simultaneously. Suppose that each part is only the size of an electron, which is the smallest particle that exists in the universe, so far as we know. Then, let us fill the entire universe (of radius 5 billion light-years) with solidly-packed sets of electrons. It can be shown that the whole universe could only contain, at the most, 10^{130} such sets of 200 solidly-packed electrons. Thus, we now are trying to visualize 10^{130} sets of 200 parts each, and trying to arrange *only one* set into the correct alignment by chance, just *once* in ten billion years, anywhere in the universe.

Suppose also that we invent a machine capable of making not one trial per second, but a billion-billion different trials every second, on every one of the 10^{130} sets. Surely this is the maximum number of possible trials that anyone could possibly conceive as ever being made on this type of situation. This would permit a total of $(10^{130})(10^{18})(10^{18})$, or 10^{166}, trials to be made.

Still, after all this, the chance that one of these 10^{166} trials would give the right result and make the system work is only one out of $10^{375-166}$, or 1 in 10^{209}. In other words, the idea that a system of 200 parts could be arranged by chance into the correct order is absolutely absurd!

Most systems, of course, including all living organisms, are far more complex than a mere 200 parts. The cerebral cortex in the human brain, for example, contains over 10 billion cells, all arranged in proper order, and each of these cells is itself infinitely complex!

The obvious conclusion is that complex, ordered structures of any kind (and the world is full of them) simply could never have happened by chance. Disorder never spontaneously turns into order. Organization requires an organizer. The infinite array of complex effects seen in the universe must have been produced by an adequate cause. An adequate Cause is God, the Creator, and nothing less!

The modern Darwinian evolutionist thinks he has a naturalistic explanation for all this, of course. The magic formula which transforms electrons into living cells, and frogs into princes, is "random mutation and natural selection," and the magic wand which makes it work is "billions of years."

As the modern leader of evolutionary thought, Sir Julian Huxley, has said:[1]

> "Darwinism removed the whole idea of God as the creator of organisms from the sphere of rational discussion. Darwin pointed out that no supernatural designer was needed; since natural selection could account for any known form of life, there was no room for a supernatural agency in its evolution."

Similarly, but more recently, Francisco J. Ayala, of Rockefeller University, has insisted on this perspective as follows:

> "Darwin substituted a scientific teleology for a theological one. The teleology of nature could now be explained, at least in principle, as the result of natural laws manifested in natural processes, without recourse to an external Creator or to spiritual or non-material forces."[2]

As we shall see in later chapters, however, Darwinism is not really a "scientific teleology" at all. Random mutations necessarily generate, not order, but disorder, and natural selection at best constitutes a screening mechanism for sieving out the disorganized misfits produced by mutational pressures, and thus for conserving the original complex systems already present. Furthermore, the longer the period of time available for such pressures to operate, the more likely it is that they will tend to overcome the selection process and drag the entire biosphere downhill to lower levels of order!

The tenet of neo-Darwinism—that random mutation combined with natural selection eliminates the need for God—is premature, to say the least. It has already been shown that complex ordered structures could never be produced by "random selection" of their parts. Nevertheless, evolutionists insist that "selection" can somehow affect the "random" nature of the process. Even though genetic mutation is a random process, they feel that natural selection can so efficiently sieve out the "good" mutations that its randomness is gradually converted into increased order, even without an intelligent organizer to control it.

But this is asking far too much of such an impersonal, unintelligent, static phenomenon as natural selection. At the very best, natural selection cannot produce the mutations—it does not energize or organize anything itself. All it can do is "decide" whether or not a combination of parts presented to it by the *random* mutation process is more ordered or less ordered than its non-mutated predecessor combination; and we have already seen that a *random* process could never produce an ordered structure for selection to "select"—even such a relatively simple structure as one containing only 200 components.

At this point the evolutionist might object that he is being misunderstood. He does not propose that an ordered structure should be suddenly organized from its 200 separate parts all simultaneously. Rather, the process works gradually, part upon part, slowly over long ages. Only one part is added at a time.

However, a little consideration will show that this only makes matters worse. The same selection process has to take place over and over again, and each time against greater odds than the time before.

That is, when the structure advances from one to two parts, it has two "choices" as to alignment, and therefore a 1 in 2 chance of success. When it goes from two to three parts, it has six choices, and therefore a 1 in 6 chance, and so on. If it goes all the way to 200 parts, its final advance has, as calculated earlier, only a 1 in 10^{375} chance. Each step in the chain has to keep "trying" until it hits on the right combination at each step before it can go on to the next one. Thus, the probability of developing an ordered structure of 200 parts by this step-by-step mutation-selection technique is only 1 out of the number represented by the series 2! + 3! + 4! + ... + 199! + 200!. This number is obviously far larger than the 200! (or 10^{375}) improbability of the system arising all at once. The evolutionist should have left well enough alone!

Now, admittedly, the above analysis assumes that each successive step must in effect "start from scratch", and this isn't really fair. The evolutionist does not visualize all parts being completely reshuffled at each step. On the other hand, it must also be remembered that in every living organized system there is an intricate inter-dependence of all parts upon each other. The elevation of an nth degree ordered system to an $(n + 1)^{th}$ degree ordered system is certainly far more involved than a mere linking of the new part on to the previous structure unchanged. And there is also the question of where the new part comes from in the first place. A mutation may cause a change in an existing part, but how does it create a new part to add to the system? Furthermore, if there is to be a change, what is to prevent the system from going downhill to less order instead of uphill to higher order?

As a matter of fact, if mutations constitute the mechanism for producing this supposed increasing complexity, it is far more likely that the system will become less ordered with each change in order, instead of more ordered. All evolutionary geneticists agree that the great majority of mutations are harmful, with not more than one in a thousand being really beneficial (actually there are probably *no* true mutations that have ever been demonstrated to be permanently helpful in the natural environment).

However, let us give the evolutionary process the maximum possible benefit of the doubt and assume that each successive step has a 50:50 chance of success. That is, for a given structure, the probability that the next change will be an addition of order is assumed to be exactly the same as that it will be a decrease of order. The probability of success at each step is 1 out of 2.

There are 200 steps, of course, to be made to arrive at an integrated 200-component system. If any one of these steps fails (that is, a "lethal" or otherwise "harmful" mutation) then of course the evolutionary process in that particular system either stops altogether or goes backward.

All 200 steps must succeed and the probability of success at each step is ½. Elementary statistical theory shows that the probability of success of the whole chain of steps is the product of the probabilities for each step. That is, the probability that a 200-step evolutionary chain can succeed is only one out of $(½)^{200}$, or 1 in 2^{200}, even assuming beneficial mutations are as frequent as harmful mutations.

This number is equal to a number represented by 1 with 60 zeros, or 10^{60}, or a quadrillion quadrillion quadrillion quadrillion. Not a very likely sequence!

Therefore, even this very simple evolutionary sequence of 200 steps is for all practical purposes impossible. An ordered system can, by no means mathematically conceivable, ever arise by a random process from non-ordered components, even if a screening mechanism such as natural selection is available to conserve its acceptable products.

The evolutionist may still offer one faint objection, saying that even though one given system only has a 1 in 10^{60} chance of evolutionary success, there must be at least *some* systems in the world that make the grade.

So, let's go this one more mile with him. The surface area of the earth contains about 3×10^{13} sq. ft. Assume that each part of the system is actually a living cell, and the entire surface of the earth is covered with living cells stacked one foot deep. There would be about 10^{13} cells then piled on top of each square foot, and the earth's surface would hold about 3×10^{26} such cells. If all of these are operated in this evolutionary process simultaneously, there would be, therefore, about 10^{24} systems of 200 parts each available in the entire earth. Since each of them has one chance in 10^{60} of evolutionary success, the chance that one out of all the 10^{24} sets would succeed is thereby reduced to 1 in 10^{36}. This is still an impossibly large number.

However, as each set fails, then let a fresh set come in and try again. Suppose each step takes one-half second, so that the 200 steps in each set would take 100 seconds. Then, in the 10^{18} seconds of astronomic time, each of the 10^{24} sets could have 10^{16} tries. Thus, a total of 10^{40} attempts could be made in all time in all the world to evolve a structure of 200 parts. The probability that one of them would ever succeed anywhere is still only 1 in 10^{20}, or one chance out of ten billion times ten billion.

And this is only one very simple structure! The world is full of great numbers of far more complicated structures and systems than this.

We conclude that evolution by any kind of chance

process, even with natural selection operating, is mathematical and scientific nonsense.

The argument from design, therefore, has not at all been refuted by natural selection theory, but is actually stronger than ever. The innumerable and marvelous structures and systems of the cosmos, and their intricate adaptations to each other constitute a vast complex of intelligible order for which creative forethought and design can be the only rational explanation. It is pointless to give examples because literally every system of any kind in the whole cosmos is itself a marvelous model of intricate structure and complex planning.

The great Designer who created this wonderful world can be none other than the God of the Bible—omnipotent, omnipresent, omniscient, holy—yet also personal, loving, and gracious. The Cause of all the phenomena of the universe must encompass at least all their own characteristics.

> "He that planted the ear, shall he not hear? he that formed the eye, shall he not see?—he that teacheth man knowledge, shall he not know?" (Psalm 94:9,10).

The finely-balanced structure of the earth's hydrosphere, atmosphere, and lithosphere are stressed in the rhetorical question of Isaiah 40:12:

> "Who hath measured the waters in the hollow of his hand, and meted out heaven with the span, and comprehended the dust of the earth in a measure, and weighed the mountains in scales, and the hills in a balance?"

As if in answer, the prophet replies by stressing God's omnipotence and omniscience:

> "Lift up your eyes on high, and behold who hath created these things, that bringeth out their host by number: he calleth them all by names by the

> greatness of his might, for that he is strong in power; not one faileth" (40:26).

The tremendous evidence of design and order in nature encourages us to testify, with the Psalmist:

> "O Lord, how manifold are thy works! in wisdom hast thou made them all: the earth is full of thy riches" (Psalm 104:24).

---

[1]*Issues in Evolution* (Sol Tax, Ed., University of Chicago Press, 1960) p. 45

[2]"Teleological Explanations in Evolutionary Biology" *Philosophy of Science,* Vol. 37, March 1970, p. 2

CHAPTER II
THE RIGOROUS WAYS OF NATURAL LAW

The world that God created is not a dead, static, unchanging thing. Rather it teems with activity, with things happening, with life. Not only does the creation exhibit an infinite variety of marvelously designed structures and relationships, as discussed in Chapter I, but also there is an unlimited complexity of inter-actions between these systems.

These inter-actions are called processes, and the study of these processes is the function of scientists. Because of the great number of different systems and processes, it has been necessary for science to divide and subdivide itself over and over again. Not only are there physicists and chemists, biologists and geologists, and other such basic scientists, but also physical chemists, organic chemists, nuclear physicists, classical physicists, and numerous other specialists within these basic disciplines. Many fields of science which once were special emphases in physics or one of the broad sciences have developed into independent branches of their own—sciences such as meteorology, hydrology, ecology, metallurgy, paleontology, and many others.

All of which points up both the extreme breadth and complexity of science and also the impossibility of any one scientist ever becoming a real first-hand authority in more than a very restricted scientific specialty. Furthermore, scientists as individuals are real people and therefore subject to the same conceits, prejudices, and other weaknesses as non-scientists. Scientists should accordingly be very cautious about making broad pronouncements on sociological or religious matters in the name of "science," and laymen should be carefully skeptical about such pronouncements when they do make them.

In spite of the great number and variety of scientific processes, there are two statements that can be made about all of them without exception. These are:

1. All processes involve interchanges and conversions of an entity called *energy,* with the total energy remaining constant. Scientifically this is called the *law of conservation of energy,* or the First Law of Thermodynamics.
2. All processes manifest a tendency toward decay and disintegration, with a net increase in what is called the *entropy,* or state of randomness or disorder, of the system. This is called the Second Law of Thermodynamics.

Thus all the processes of nature are fundamentally

processes of quantitative *conservation* and qualitative *disintegration.* These two laws, accepted by all scientists as the most universally-applicable principles which science has been able to discover, were only recognized about a hundred years ago.

However, these basic principles have been in the pages of the Bible for thousands of years, though not expressed in modern scientific terminology. The conservation principle is clearly set forth by the fact of a *completed* creation which is now being *sustained* by its Creator.

Colossians 1:16,17, for example, indicates both aspects of this truth. "By Him were all things created . . . and by Him all things consist." Note that "created" is in the past tense. The Scripture does *not* say: "By Him are all things being created." Thus, creation is not going on at present. The word "consist" is a translation of the Greek word from which we get our English word "sustain." Thus, the verse says in effect that "by Him all things are sustained." By the Lord Jesus Christ, all things—all systems and structures, all kinds of organisms and relationships—were created once for all in the past and are now being conserved.

This same principle—that nothing is now being created or destroyed—is also implied in many other passages. Examples include Hebrews 1:2,3: ". . . He made the worlds . . . upholding all things by the word of His power"; II Peter 3:5,7: "by the word of God the heavens were of old and the earth . . . the heavens and the earth which are now, by the same word are kept in store"; Psalm 148:5,6: ". . . He commanded, and they were created. He hath also established them for ever and ever"; Isaiah 40:26: ". . . He hath created these things, . . . for that He is strong in power; not one faileth"; Nehemiah 9:6: ". . . thou hast made heaven, the heaven of heavens, with all their host, the earth, and all things that are therein, the seas, and all that is therein, and thou preservest them all."

The first chapter of Genesis describes this creation, and it should be stressed as strongly as possible that it is only in the Bible that we can possibly obtain any information about the methods of creation, the order of creation, the duration of creation, or any of the other details of creation. Since, according to both Scripture and the First Law of science, nothing is now being created, therefore the scientific study of present processes can reveal nothing about creation except that it must have taken place. This is the most fundamental fallacy in the evolutionary theory. Evolution assumes that these present processes are the same processes by which all things have developed from primeval chaos into their present complexity. Both the Word of God and the First Law of science say otherwise.

At the end of the account of creation, the record is very explicit and definite: "Thus the heavens and the earth were finished, and all the host of them. And on the seventh day God ended His work which He had made, and He rested on the seventh day from all His work which He had made. And God blessed the seventh day, and sanctified it: because that in it He had rested from all His work which God created and made" (Genesis 2:1-3).

This summary is very clear in its insistence that whatever methods God used in creating and making all things–including man himself–He stopped using. The present work of providence–of providing for the conservation and sustenance of all the basic entities He had created–is of a different order altogether from His work of creation.

Superimposed on the conservation principle, however, is the decay principle. The Second Law of Thermodynamics, no less than the First Law, is a universal law governing all processes. Although energy is never destroyed, it continually becomes less available for further work. Everything tends to wear out, to run down, to disintegrate, and ultimately to die. All processes, by definition, involve change—but the change is not a change in the upward direction, such as the evolutionists assume.

Somehow it seems contrary to the nature and purposes of God that He would create a universe in which decay and death constitute one of the basic principles. He is a God of grace and power; as a gracious and merciful God surely He would not build such a principle into His creation if He could help it. Since He is also omnipotent, He certainly did not have to do it this way.

Why, then, do "we know that the whole creation groaneth and travaileth in pain together until now." (Romans 8:22)? There is no doubt that the Second Law is a universal law, in both the physical and biological worlds, so far as science can determine. Left to itself, everything collapses, deteriorates, grows old, and dies, sooner or later. Eventually the whole universe seems destined to die, when all the energy of the sun and stars will have been degraded to uniformly-dispersed low-level heat energy, no longer capable of being converted into useful work.

Is this what God intended, when He finished His creation and pronounced it all "very good" (Genesis 1:31)? Obviously not; God is not capricious, and we can be absolutely sure He will accomplish His good purpose in creation.

The answer can be only that the Second Law is a sort of intruder into the divine economy, not a part of either the original creation or God's plan for His eternal kingdom. God's description of the entire creation as "very good" must tell us that at that time there was no disorder, no deterioration, no groaning and travailing, no suffering, and, above all, no death in the whole universe, "the heavens and the earth, and all the host of them" (Genesis 2:1).

The imposition of the principle of decay and death on the original creation was the result of man's sin. God had to bring the Curse upon both man and his dominion because of man's rebellion against his Creator. "Cursed is the ground," God told Adam (Genesis 3:17). The very ground out of which Adam's body had been constructed, the dust of the earth, the basic elements of the creation, were thus brought into the "bondage of decay" (Romans 8:21). "Yea, all of them shall wax old, as a garment." (Psalm 102:26).

The Curse is not permanent and irrevocable, however, but only remedial and disciplinary. "The creation itself also shall be delivered from the bondage of decay into the glorious liberty of the children of God" (Romans 8:21). "There shall be no more curse" (Revelation 22:3). "We, according to His promise, look for new heavens and a new earth, wherein dwelleth righteousness" (II Peter 3:13).

At the same time God pronounced the Curse, He also promised the coming Redeemer, the seed of the woman (Genesis 3:15) who would someday provide salvation, both for individual souls and for the whole creation.

In the meantime, however, in so far as our present study is concerned, we can see now that the two basic laws of science, the First and Second Laws of Thermodynamics, are merely man's scientific statements of the two revealed facts of: (1) a creation originally completed and now sustained by God's power, and (2) the curse of decay and death, superimposed on the creation by its Creator because of man's sin.

Rather than science disproving the Scriptures, as many allege, the two best-proved, most universal laws of science, within which all processes must operate, were anticipated and recorded in Scripture thousands of years before man discovered them! Furthermore these two laws give a clear and sure testimony to the fact of creation, and therefore of a Creator. The Second Law demonstrates that there must have been a beginning, or otherwise the universe would already be dead. The First Law demonstrates that the universe could not have begun itself, since none of its processes creates anything. Thus the only logical conclusion is that "in the beginning God created the heaven and the earth."

Furthermore they tell us plainly that the processes of nature are conservative and disintegrative, not innovative and integrative, as the evolutionist maintains. There *is* a universal process of change operating in the world, and often the evolutionist tries to define evolution merely as "change." However, he really means a *directional* change, whereby molecules slowly change into men, over aeons of time. The real law of change, however, is one of *decay,* not of growth, a change "down" instead of a change "up." Thus the laws of thermodynamics sharply conflict with the philosophy of evolution. The latter is at best a religious faith, not science.

Some writers have attempted to circumvent the witness of the Two Laws against evolution by arguing that the earth is an "open system" and the sun's energy is great enough to offset their effects. However, *all* systems in the world are "open" in some degree to the sun's energy, so this in itself is no argument.

There is a universal tendency for all systems to go from order to disorder, as stated in the Second Law, and this tendency can only be arrested and reversed under very special circumstances. We have already seen, in Chapter I, that disorder can never produce order through any kind of random process. There must be present some form of *code* or *program,* to direct the ordering process, and this code must contain at least as much "information" as is needed to provide this direction.

Furthermore, there must be present some kind of mechanism for converting the environmental energy into the energy required to produce the higher organization of the system involved. Even if there is enough energy from the sun in the environment, it will not automatically transform itself into some kind of orderly structural growth in a system. There must be an efficient and powerful energy converter present if the work is to be done.

Thus, any system that experiences even a temporary growth in order and complexity must not only be "open" to the sun's energy, but must also contain a "program" to direct the growth and a "mechanism" to energize the growth. Otherwise the system will merely disintegrate and die, regardless of the sun's energy.

Now the imagined age-long evolutionary growth of the whole world of organisms has neither program to direct it nor mechanism to empower it. Neither mutation nor natural selection is a program, and neither mutation nor natural selection is an energy conversion device. Neither one is either one. To offset the Second Law, and produce true evolution, evolutionists need still to find a directing code and an enabling mechanism, and neither of these has yet been disovered.

CHAPTER III
THE TERRIBLE DEATH OF THE ANCIENT WORLD

When confronted with the fact that no evolution from one basic kind to another has ever been observed to take place in the historic period and that, in fact, the basic nature of natural processes (as specified by the First and Second Laws of Thermodynamics) seems to preclude the very possibility of significant evolutionary changes of this type, the evolutionist must eventually fall back on the fossil record. He regards this record as firm proof of evolution, documenting the gradual increase in complexity of the organic world with the passage of aeons of geologic time.

Most of the surface of the earth's crust is covered with great thicknesses, sometimes up to several miles deep, of sediments and sedimentary rocks, and these normally contain fossil remnants of plants and animals which once lived on the earth. The deeper rocks normally contain simpler fossils than the shallower rocks (though there are many significant exceptions to this rule). In fact, the very concept of the geologic ages is based on this assumed gradation of fossils. The various individual geologic ages are supposedly recognized and identified by the evolutionary stage of the fossil assemblages found in them.

However, there are many scientific difficulties with this interpretation of the fossils. The very existence of fossils in any significant size and number seems to require rapid processes of sedimentary deposition, burial, compaction, and lithification. Otherwise, normal decay processes would soon destroy and dissipate such organic remains.

Furthermore, the fossil record does not show a continuous evolutionary progression at all, as the theory requires. The same great gaps between the major kinds of plants and animals that exist in the present world are also found in the fossil world. Of course many animals that once lived have become extinct (such as the dinosaurs); but extinction is not evolution!

In addition, as mentioned above, there are numerous localities around the world where supposedly older and simpler fossils have been deposited in layers vertically above layers containing "younger," more complex, fossils. One famous area of this type is in Glacier National Park where a vast block of Precambrian limestone, supposedly nearly a billion years old, is found resting on top of Cretaceous shales, presumably only (!) a hundred million years old. The remarkable thing about this is the fact that this out-of-order block of limestone is perhaps 350 miles long, 35 miles wide and 6 miles thick, with every appearance of having been laid down by normal sedimentary processes on top of the "younger" shale beds below.

In fact, there seems to be no known present physical process that could produce such a gigantic "overthrust" as this. If indeed it could ever be demonstrated by physical proof that the order of the strata had really been inverted by actual movements after their original deposition, it would prove also that forces entirely different from anything now in existence must have been acting in the past. The standard geological dogma of uniformitarianism–namely, that processes of the same kind and intensity as at present are able to explain all past phenomena–is thereby proved false either way.

This is only one example, out of scores that could be mentioned. Furthermore, there are even greater numbers of examples of missing "ages" in localities everywhere. It is not too much to say that, in some location or other, one could find practically any sequence whatever of the so-called geologic ages.

Obviously there is something basically wrong with the evolutionary interpretation of the fossil record. Is there an alternate explanation, which would not confront such problems as these, as well as many others which could be mentioned? Does the Bible give any light on this?

Most creationists are convinced that the key to the real understanding of the fossil-bearing sedimentary rocks is nothing less than the great Flood of the days of Noah. The fossils speak, not of the gradual evolution of life on earth over vast ages, but rather of the sudden extinction of life, all over the world, in one age.

According to the Bible, there was no death before Adam sinned. "As by one man, sin entered into the world, and death by sin" (Romans 5:12). That is, death came into God's perfect world only when sin came into the world through man. The fossils, however, speak of death, and on a gigantic scale!

There was never such an opportunity for production of fossils as in the great Flood. Neither was there ever, before or since, such an opportunity for the formation of vast beds of sediment and for their rapid conversion into sedimentary rock, as in the great Flood. "And every living substance was destroyed which was upon the face of the ground, both man, and cattle, and the creeping things, and the fowl of the heaven; and they were destroyed from the earth; and Noah only remained alive, and they that were with him in the ark" (Genesis 7:23).

The Biblical record of the Flood is not merely an exaggerated description of a local flood, as many have thought, but instead describes a worldwide aquaeous cataclysm, "Whereby the world that then was, being overflowed with water, perished" (II Peter 3:6).

God told Noah: "The end of all flesh is come before me; for the earth is filled with violence through them; and behold, I will destroy them with the earth" (Genesis 6:13). Obviously, if all men on the earth were to be destroyed by a flood, then the flood would have to cover the earth, because man had *filled the earth* with violence. In fact, God said explicitly that the *earth* would be destroyed, along with the men on the earth.

The construction of Noah's ark makes no sense at all, of course, if the Flood were only a local flood. Such a huge structure, easily capable of accommodating two of every known species of dry-land animal, living or extinct, was absurdly unnecessary if the birds and animals could simply have migrated out of the flood plain.

When the Flood came, it soon covered all the mountains. As a matter of fact, it was eight months after the Flood began before the waters went down enough so that "the tops of the mountains were seen" (Genesis 8:5). The ark in the meantime had rested "upon the mountains of Ararat" (Genesis 8:4).

Mount Ararat at present is 17,000 feet high, and there are also many other high mountains in the region where the post-diluvian dispersion began. Obviously, therefore, the Flood would at least, according to the Bible, have covered these mountains. But a flood which can cover a 17,000-ft. mountain for eight months is not a local flood!

Furthermore, after the Flood, God promised "neither shall there any more be a flood to destroy the earth" (Genesis 9:11). If the Flood had been only a local flood, God has not kept His promise, because there have been many other disastrous local floods since that time, in all parts of the earth.

Finally, the Flood is referred to in many later parts of the Bible, always in such a way as to indicate that it was worldwide (Psalm 104:6–9; Isaiah 54:9; Hebrews 11:7; II Peter 2:5; etc.). The Lord Jesus Christ Himself said: "And as it was in the days of Noah, so shall it be also in the days of the Son of man, . . . the flood came and destroyed them all" (Luke 17:26; see also Matthew 24:37–39). Those people who advocate the local flood theory obviously do not take the words of Scripture, even those of Christ, very seriously. In Appendix A will be found a list of 96 arguments for a worldwide flood, 64 Biblical reasons and 32 non-Biblical reasons.

It should be noted briefly that some Bible students, seeking to find some means of explaining the fossils without having to attribute them to the Flood of Genesis, have suggested that they may have been formed in a pre-Adamic cataclysm which left the earth "without form and void," as described in Genesis 1:2.

This suggestion, however, contradicts the explicit Biblical teaching that there was no death in the world prior to Adam's sin (Romans 5:12; I Corinthians 15:21), as well as God's evaluation of the entire universe as "very good" (Genesis 1:31–2:3) at the end of the six days of creation.

The pre-Adamic cataclysm theory is illogical and unscientific as well. Any cataclysm of such devastating nature as to leave the entire globe shrouded in absolute "darkness on the face of the deep" (Genesis 1:2) must have been nothing less than a shattering worldwide nuclear or volcanic explosion of some kind. Whatever its exact nature may have been, such a thermal holocaust would unquestionably have utterly annihilated any evidence of previous living organisms. It could never have produced the finely-stratified sedimentary beds of the earth's crust, with their abundance of finely-structured fossiliferous contents.

Consequently, the fossil record must be attributed to the Noahic Deluge and perhaps, in lesser measure, to subsequent related regional disturbances stemming from the after-effects of the Flood. Although there are still many unsolved geological problems in connection with this Biblical interpretation of the fossils, these problems are not nearly so difficult to interpret as those confronting the evolutionary geologist. Many creationist scientists today are actively doing research on these problems and the scientific basis of so-called "flood geology" is becoming stronger all the time.

The evolutionary interpretation of the fossil record is based on the assumption that the simpler fossils are found in the older rocks. However, it is also true that rocks are dated primarily by the fossils they contain.

The main proof of evolution (the fossil sequence) is derived from the relative age of the rocks, which is determined on the basis of the assumed evolutionary sequence of the fossils. *One can prove* ANYTHING *if he starts with his conclusion and then reasons in a circle.*

Furthermore, as noted before, there are many local exceptions to the standard fossil order found in the actual rocks. Outlandish physical mechanisms with no parallel in the modern world have to be invented to explain them, building hypothesis on hypothesis on hypothesis.

There is, in fact, no way by which one geologic "age" can be distinguished from another except by its fossils. There are no physical differences in structure or appearance. Radiometric "ages" (as discussed in Chapter VIII) are extremely variable and are only accepted if they happen to coincide with the assumed fossil age.

Furthermore, there is no dividing line in the rocks between the ages. As far as physical aspects are concerned, each "age" merges imperceptibly into the next.

The fossils themselves, of course, all speak of sudden death and rapid burial—otherwise they would quickly have been destroyed by bacteria, scavengers, and other agents of decay. Each fossil deposit, therefore—and it is these which "date" the rocks—is a witness to catastrophic deposition of its own beds.

Since the rocks, therefore, everywhere speak of

catastrophism, rather than uniformitarianism, and since there is no way by which one "age" is distinguished from another and no evident time lapse from one rock system to another, it seems reasonable to think of the entire fossil-bearing assemblage of strata as having been formed in only one great worldwide catastrophe. This theory of course allows a tremendous complexity of localized phenomena and the formation of a wide variety of stratigraphic sequences in different parts of the world. It involves a global hydraulic and sedimentary cataclysm, accompanied by great volcanic and tectonic movements and then followed by a drastic change in climate and continental glaciation.

The general order of fossils, from simple on the bottom to complex on top, is exactly what would be expected in such a cataclysm. The different geologic "ages" are actually different ecologic zones in the one antediluvian age. Simple marine organisms tend to be found buried at the lowest elevations in the geologic column for the simple reason that they lived at the lowest elevations. Birds and mammals are found in the more "recent" geologic epochs mainly because they lived at higher elevations and had greater mobility.

Though there are unsettled problems in this flood theory of geology, they are not nearly so serious, nor do they require nearly so many auxiliary hypotheses, as in the evolutionary theory.

The cause of the deluge was the concurrent release of the "waters above the firmament" (Genesis 1:7) through the "sluiceways of the heavens" and the waters locked under the crust through "the fountains of the great deep" (Genesis 7:11;8:2). Prior to the deluge, the waters above the "firmament" (literally "expanse," referring to the atmosphere) had apparently been in the form of a vast canopy of water vapor, invisible and, therefore, translucent to the light of the heavenly bodies (Genesis 1:16), but producing a marvelous worldwide "greenhouse effect". The climate was warm and mild everywhere, thus inhibiting air mass movements of a global nature and precluding rain as we know it now (Genesis 2:5;9:13). It also effectively filtered out radiations from space which now are known to have an accelerating effect on the aging process, thus probably being a chief agent in the maintenance of patriarchal longevity before the flood (Genesis 5:5,27).

The warm subcrustal waters emerged under controlled conditions through great artesian springs (Genesis 2:10-14) to supply the antediluvian rivers. There were no great deep and wide oceans, but rather a network of epi-continental seas (Genesis 1:10). There were no deserts or ice-caps and the greenhouse climate supported lush vegetation and animal life on the lands and abundant marine life in the seas (Genesis 1:20).

The coming of the great flood changed all this. The mild topography and pleasant climate of the old world became the rugged terrains, vast oceans, and violent climates of the new world. ". . . the waters stood above the mountains. At thy rebuke they fled, at the voice of thy thunder they hasted away. (The mountains rose up, the valleys sank down unto the place which thou hast founded for them.) Thou has set a bound that they may not pass over; that they turn not again to cover the earth" (Psalm 104:6-9).

When one realizes that the fossils do not represent a long-drawn-out dimly-understood evolutionary history of the earth but rather a graphic record in stone of a former world much like our own, with which we are directly connected through our great ancestor Noah, then the study of fossils and geology becomes dynamic and vitally relevant to our world today. God has preserved all over the earth, for everyone to see as long as the present world endures, this tremendous monument to His sovereign control of the created universe. Just as He was able to destroy the rebellious "world that then was" by the waters of the Flood, so He is still able to see that "the heavens and the earth which are now . . . are kept in store, reserved unto fire against the day of judgment and perdition of ungodly men" (II Peter 3:6,7).

The great coal beds of the world are recognized, not as the accumulations of age after age of peat-bog growth, but rather as the transported and metamorphosed remains of the extensive and luxuriant vegetation of the antediluvian world. The oil reservoirs are the traps into which the compressed and converted remains of millions of buried marine animals have migrated after burial in the subterranean upheavals of the "fountains of the great deep."

The great fossil graveyards of land vertebrates—reptiles and mammals—are recognized in terms of Biblical geology to be herds of pre-diluvian animals which were overtaken by the vast sediments propelled by the Flood waters and buried before they could escape. In some of these cases, there may also be the possibility of burial by some post-Flood regional catastrophe. Great volcanic lava flows, earth movements, violent windstorms, and other catastrophes, including the great Ice Age, were after-effects of the Flood, resulting from the global cataclysmic changes in the earth's lithosphere, hydrosphere, and atmosphere during the Flood.

The most spectacular of the fossil deposits, of course, are those of the dinosaurs. It is primarily the dinosaurs which have caused many people to think that the fossil world is so different from the modern world that it must have existed millions of years ago, long before modern plants and animals evolved.

Other than the dinosaurs, however, most of the fossil organisms have present-day descendants which are essentially still the same. One thinks of crocodiles and turtles, starfish and clams, sharks and eels, cockroaches and scorpions, bats and pelicans, opossums and coneys, and innumerable other modern animals, all of which are found in the fossils, in rocks supposedly laid down far back in geologic time.

It is not at all far-fetched, in fact, to think that the dinosaurs themselves lived contemporaneously with man in the antediluvian world. Dinosaur footprints have been found in abundance in rocks in many places, and these are occasionally found associated with what appear to be human footprints in the same rocks. One location in central Texas has yielded at least two dozen man-like tracks along with the dinosaur tracks. A number of these are in distinct trails—right foot, left foot, right foot, etc.—and some of the human-like tracks actually overlap dinosaur tracks.

Also there are places known,—in Rhodesia and Arizona, for example,—where dinosaur pictographs, drawn on cave or canyon walls by some prehistoric human artist, have been found. The evidence seems strong that the artist had somewhere seen living dinosaurs.

The universal primitive belief in dragons may be significant in this connection. How can we account for the fact that every nation seems to have traditions dealing with dragons unless their distant ancestors somehow encountered fearsome reptilian creatures around which such tales developed?

The Bible frequently mentions dragons, for that matter, and this can only mean that such creatures really existed within the period of man's history (note

Psalm 148:7, Isaiah 43:20; etc.). In fact, the account of creation may well have mentioned them. "God created great whales, and every living creature that moveth, . . . And God made the beast of the earth after his kind . . ." (Genesis 1:21,25). The Hebrew word for "whales" is *tannin,* and in certain other passages is actually translated "dragons."

The book of Job is one of the oldest in the Bible and reflects living conditions in the early centuries after the Flood. The climax of the book is when God speaks directly to Job and his friends in Job 38,39,40 and 41. God is calling attention to His great power in creating and sustaining all things (exactly the message urgently needed by the world today).

Finally He calls attention to His two greatest creations in the animal kingdom, behemoth (Job 40:15–24) and leviathan (Job 41:1–34). Most commentators today suggest behemoth is either the elephant or hippopotamus and that leviathan is the crocodile. However, the actual descriptions (and these, coming as they do from the mouth of God Himself, certainly refer to real animals) obviously do not apply to any animals known today. The most reasonable interpretation, therefore, is that they refer to extinct animals. Perhaps, then, behemoth is a land dinosaur and leviathan a marine dinosaur. Suddenly these chapters become very much alive and meaningful! These great animals were still living in Job's day, even though they may have become extinct since. The cause of the extinction of the dinosaurs is, of course, still a mystery to uniformitarian geologists even today. However, the Flood and its devastating residual effects on post-diluvian climates and food supplies constitute an adequate solution to the mystery.

In reading God's description of behemoth, one can

clearly visualize a giant brontosaur, with his long neck projecting out to eat the swamp vegetation and to wash it down with great quantities of water, and with his powerful legs and tail easily capable of demolishing his enemies with their overwhelming blows. "Behold now behemoth," God remarks, . . . "he eateth grass as an ox . . . his strength is in his loins, and his force is in the navel of his belly. He moveth his tail like a cedar." (—ever see an elephant's tail?) "His bones are as strong pieces of brass, his bones are like bars of iron" (no wonder so many fossil dinosaur bones have been preserved so long). Finally God states, "He is the chief of the ways of God" (thus the greatest animal God ever made); "He that made him can make his sword to approach unto him" (thus God Himself can destroy the dinosaurs, even though man could not). "Behold, he drinketh up a river, . . . his nose pierceth through snares."

Thus the great dinosaurs, along with all other creatures, were made by God in the six days of creation. The fossil record is a continuing testimony to God's sovereign destruction of that first creation in the great Flood.

CHAPTER IV
THE MARVELOUS NATURE OF LIVING THINGS

The marvel of life can only be explained by creation. One of the strangest phenomena of our supposedly scientific age is the insistent faith held by many *scientists* (!) that somewhere, somehow life has arisen from non-life by naturalistic evolutionary processes. Science is supposed to be based on facts and knowledge, not speculation and wishful thinking. The law of biogenesis, based on all the *observed* data of biology and chemistry, states that "life comes only from life." The doctrine of *abiogenesis,* on the other hand, teaches that certain unknown conditions in the primitive atmosphere and ocean acted upon certain mysterious chemicals existing at that time to synthesize still more complex chemicals which were able to reproduce themselves. These replicating chemicals, whatever they were, constituted the original living systems from which all living organisms later evolved.

Thus, primeval *unknown* life forms which no longer exist were derived from *unknown* chemicals by *unknown* processes which no longer operate, in an atmosphere of exotic and *unknown* composition in contact with the primitive oceanic soup of *unknown* structure! This remarkable construct is today taught as sober *science* in our public schools, in spite of the fact that there is not one single scientific observation to demonstrate that such things ever happened or even could happen.

Of course, millions of dollars in research grants have been fruitlessly spent in the search for some such evidence. Amino acids have been synthesized, genes and viruses and cells have been disassembled and then reassembled, and "proteinoid" blobs have been constructed in futile attempts to manufacture true protein molecules. But all such activities are as far from the true creation of life as a rock-pile is from the Taj Mahal.

This yearning to make something that has life is a modern equivalent of the ancient pagan idolatry. "Their idols are silver and gold, the work of men's hands. They have mouths, but they speak not: eyes have they, but they see not: They have ears, but they hear not: noses have they, but they smell not: They have hands, but they handle not: feet have they, but they walk not: neither speak they through their throat. They that make them are like unto them; so is every one that trusteth in them" (Psalm 115:4–8).

Even more foolish than man's presumptuous attempt to create anything else that has life is his belief that blind chance acting on non-living molecules could do it. Only God, "in whom is life" (John 1:4), can produce life. An effect must have an adequate Cause.

There are three specific acts of special creation recorded in Genesis One: (1) the creation of the basic inorganic components of the space-mass-time universe (Genesis 1:1); (2) the creation of living creatures, the life principle, consiousness (Genesis 1:21); (3) the creation of the image of God in man (Genesis 1:27). Each of these verses utilizes the Hebrew *bara,* which implies creation in a special, unique sense, evidently a creation utilizing no previous materials of any kind, a creation out of nothing except God's own power. Between these acts of direct creation, God's work consisted of "making" or "forming" all things into their completed structures, utilizing the basic entities He had created. However, these processes were not evolutionary processes, but rather integrative processes under the active control of God Himself, involving His own direct "handiwork" (Psalm 19:1). Both types of activity on God's part—creating and making—were accomplished and finished in the six days (note Genesis 2:3; Exodus 20:11).

Man's body itself did not involve a new creation, since it was "formed of the dust of the ground" (Genesis 2:7), the basic elements of which had already been created on the first day of creation. The same is true of the bodies of the animals (Genesis 2:19).

The same is true, evidently, of the biological life which is shared by man in common with the animals. The second act of specific creation was when "God created great whales, and every living creature that moveth" (Genesis 1:21). The word "living creature" is the Hebrew *nephesh,* which normally is translated "soul." It is the same word used later when "the Lord

God formed man of the dust of the ground, and breathed into his nostrils the breath of life, and man became a *living soul"* (Genesis 2:7). Thus, both men and animals share the created entity of God called the *nephesh,* the "living soul."

One important, almost synonymous, aspect of this life is "the breath of life." That this is also shared by animals is clear from Genesis 7:21,22, where it is said that "All in whose nostrils was the breath of life, of all that was in the dry land, died." This statement is a summary of the preceding statement that "all flesh died that moved upon the earth, both of fowl, and of cattle, and of beast, and of every creeping thing that creepeth upon the earth, and every man."

The word "breath" in the Hebrew is *ruach,* and also is the same word translated "wind" or "spirit," the context determining in each case which it should be. It involves the respiratory and circulatory apparatus, which maintains biological life in the body. In the case of man and the higher animals, the bloodstream is the medium through which the life-maintaining oxygen provided by the breathing mechanism is transmitted to all parts of the body for its necessary maintenance. "The life (i.e., *nephesh,* "soul") of the flesh is in the blood" (Leviticus 17:11; Genesis 9:4).

Both man and animals, therefore, share that creation of God called the "soul" and the "breath." Since this is a special creation, and since only God can "create" (in the Bible, the only subject ever used with the Hebrew *bara* is God), it is absolutely futile for man to think he can ever create a "soul" or a "breath of life." Even if it were conceivable that man might be able someday to design some complex chemical that can replicate itself, which he might then arbitrarily define as "living," such a replicating chemical would still not possess in any sense the "breath of life" or the factor of *consciousness* which is associated with life in the Biblical sense.

This attribute of consciousness needs to be emphasized. It is apparently centered physically in the brain which, with its fantastically complex electric circuitry and associated nervous system, is undoubtedly the most highly organized and intricately structured type of system in the universe. Its functioning, of course, depends on the blood, with its "soul," and the "breath" with its oxygen. Injury to the brain causes *un*consciousness or deficient consciousness, and perhaps biological death.

The above considerations indicate that plants do not possess life in this Biblical sense. They are merely extremely complex replicating systems of organic chemicals. It is significant that they were "brought forth" (Genesis 1:12) on the third day, prior to the first creation of "living creatures" on the fifth day.

The same is perhaps true of the simpler forms of what men have defined as the animal kingdom, though the exact dividing-line between conscious living creatures and non-conscious replicating systems is not yet clear, either from Biblical definitions or scientific study. Since plants do not have life in the Biblical sense, they cannot die in the Biblical sense. When men and animals were given instructions to eat the fruits and herbs God had created, this was therefore quite consistent with the fact that there was originally no death in the world. As we have repeatedly emphasized, death came into the world only when "by one man sin entered into the world" (Romans 5:12).

Death, however, is not necessarily cessation of existence. The "soul," as well as "spirit" (same word as "breath" in both Hebrew and Greek), can exist apart from the body, since they involve a special creation of God. Though the brain may become unconscious or die physically as a result of injury or disease, somehow the consciousness of the soul and spirit can survive the unconsciousness of the brain. In fact, even when man is unconscious outwardly (through sleep or anaesthesia or brain injury) experience indicates that he is still conscious in some other sense—call it "dreaming," perhaps.

Furthermore, there is an abundance of evidence in the Bible, as well as in the phenomena of spiritism and occultism (not all of which are fraudulent or imaginary) that there do exist in the world intelligent "spirits" of some kind who are not *embodied* spirits. The "breath" can exist apart from the nostrils and respiratory system and the "soul" can exist apart from the blood, though for this present world they are normally so embodied.

There is obviously much that we do not yet understand about these matters. Science is as yet, and perhaps always will be, incapable of dealing with the phenomena associated with the soul and spirit, especially after death, or with other created spirits, such as angels or demons. It can only deal with the physical realm and those phenomena of the biological and psychological realms which are directly connected for the present with the physico-chemical structures through which they function.

As far as the Scriptures are concerned, as we have seen, both man and animals have body, soul, and breath, in their present earthly lives. There are many similarities between man and animals, in terms of both structure and function, and it is perhaps understandable that evolutionary relationships might be suggested by non-theists to explain those similarities.

However, in addition to the Biblical history of their actual distinct creative origins, there are three essential differences revealed in Scripture between man and the animals. One of these is that each man's soul-spirit complex survives death; that of the individual animals does not. Similarly, only man's body will be raised from the dead and made immortal. The reason for these two differences is found in the third difference—namely, that man was created in the image of God.

This was the third of the special creative acts of God mentioned in Genesis One: "So God created man in His own image, in the image of God created He him; male and female created He them" (Genesis 1:27). God also had said: "Let us *make* (Hebrew *asah*) man in our image, after our likeness" (Genesis 1:26), just as He had also "*made* the beast of the earth" (Genesis 1:25) and other living creatures that He had "created" (Genesis 1:21). There is evidently an almost-synonymous usage of the two verbs "create" and "make," when God is the subject of either, the one emphasizing an initial creation out of nothing except God's omnipotence, the other a rapid formation of such character and complexity as none but God could accomplish. When God "*made* man in His image and likeness," He utilized entities and attributes which He already had created and which He foreknew would one day be incarnate in His Son. When he "*created* man in His own image," He called into being a distinct and eternal individual personality, capable of intercommunion and fellowship with Himself.

When man was then told to "Be fruitful, and multiply, and fill (not "replenish," as the Hebrew *male* is incorrectly translated in the King James Version) the earth" (Genesis 1:28), the process of reproduction of course was essentially the same as that for the animals. Physical attributes, as well as biological, can be transmitted from parents to children by definite genetic laws. However, for each new person so generated, there is also a special creation which takes place, the "image of God," a unique and eternal personality, capable of fellowship with his Creator. This is not true in the case of animals, whose physical and biologic characteristics are purely the result of heredity and environment.

This process of reproduction is itself a marvelous evidence of God's wisdom and power. Man has been able to learn much about it and even to control it to a limited degree. That is, through processes of controlled breeding and selection, he can develop characteristics in plants and animals which he considers desirable for his own uses. In a sense, he is exercising in this way his God-given ability to "have dominion" over all other creatures (Genesis 1:28), though he has no doubt often misused this authority.

However, he can do this only within narrow limits. Even the wizardry of a Luther Burbank cannot make "the fig tree, my brethren, bear olive berries; either a vine, figs" (James 3:12). All the words written about evolution have never made it happen. Like always begets like; one basic kind may proliferate into many varieties, but it never changes into another basic kind.

This is always true in the real world, as distinct from the hypothetical world of evolutionary speculation and the similar imaginary world of children's fairy tales, where pumpkins turn into coaches and mice into men. A species of white moth may change into a gray moth, but never into a praying mantis. A long-legged sheep may mutate into a short-legged sheep, but never will it become a donkey. The fruit-fly, *Drosophila,* has been bred in the laboratory for a thousand successive generations and continual radiational bombardment of the unfortunate insect has produced a great variety of mutational deformities on it but it is still a fruit-fly.

One would think a thousand generations would

suffice to demonstrate some kind of evolutionary development, but none has yet appeared. The evolutionist needs still more time, and he has faith that, given *enough* time, the kind will itself evolve into a different kind. This faith is not *science,* of course, but blind faith, without any foundation in experiment or observation at all, and indeed actually contrary to all evidence, as well as the basic laws of thermodynamics.

As already noted, he does not find the supposed transitional forms in the fossil record either, but only great gaps between the kinds. This he then must attribute to periods of "explosive evolution" which occurred so rapidly they left no trace in the fossils. Again he must walk by faith, not by sight! Evolution occurs so slowly in the present world we cannot detect it and so rapidly in the fossil world we cannot detect it. "Therefore," he says triumphantly, "evolution is a fact, and everyone must believe it!" Evolutionary logic is at least an interesting study in programmed learning.

The actual fact of variation within permanently fixed kinds is, of course, clearly the teaching of Scripture. Ten times in the first chapter of Genesis, we are told that the created entities, both plant and animal were to bring forth after their own kinds (Genesis 1:11,12,21,24,25), and not after some other kinds.

This is also taught in the New Testament, in I Corinthians 15:38,39. "But God giveth it a body as it hath pleased him, and to every seed his own body. All flesh is not the same flesh: but there is one kind of flesh of men, another flesh of beasts, another of fishes, and another of birds."

"To every seed his own body." The unique structure of the genetic mechanism for each kind is now understood in terms of the *genetic code.* The transmission of the hereditary "information" from parents to progeny is evidently the function of the D.N.A. molecule, structured in the intricate double-helical coiling of its components. The information encoded is such as always to assure that the progeny will be of the same kind as the progenitors, though with an abundance of individual variation possible and necessary. When something goes wrong in this complex system (i.e., "mutation"), there is a deficient transmission of the information and the result becomes, to some degree, a disordered structure destined probably to be eliminated by natural selection.

But this is exceptional. Mutations are rare and, for the most part, the process of reproduction is marvelously efficient and wonderful. "I will praise thee; for I am fearfully and wonderfully made: marvelous are thy works; and that my soul knoweth right well. My substance was not hid from thee, when I was made in secret, and curiously wrought (literally "embroidered," perhaps an intimation of the intricate weaving of the D.N.A. molecular structure) in the lowest (or "hidden") parts of the earth. Thine eyes did see my 'substance, yet being unperfect' (this phrase is one Hebrew word that could perhaps be translated "embryo"); and in thy book all my members were written, which in continuance were fashioned, when as yet there was none of them" (Psalm 139:14–16).

CHAPTER V
THE EARLY HISTORY OF ALL MANKIND

We have already seen that man, according to the Bible, is a special creation, entirely unrelated to the animals by any kind of evolutionary connection. Even if anthropologists were able to produce fossils of creatures intermediate between men and apes (and no such fossils have yet been discovered) it still would not prove human evolution. An extinct ape could have certain man-like features and still be an ape, even as true men might possess certain simian-like features and still be men.

The uniqueness of man lies not in his physiology, which has many similarities to that of the animals, nor even in his conscious intelligence, which is shared at least to some extent with the animals, but in his spiritual nature, the implanted "image of God," his capacity for abstract thought, his awareness of esthetic and ethical values and, above all, his capability for personal fellowship with his Creator.

This unique nature of man, with all its varied cultural and social implications, was specially created by God on the sixth and last day of creation (Genesis 1:26,27). The second chapter of Genesis gives further details concerning the events of this sixth day. "Adam was first formed, then Eve" (I Timothy 2:13). In the interim between the formation of Adam's body and that of Eve, God taught Adam a very basic lesson which modern man has largely forgotten. "Out of the ground, the Lord God *had* formed (a more precise translation, in context, than the King James "formed") every beast of the field, and every fowl of the air, and brought them unto Adam to see what he would call them" (Genesis 2:19).

This procedure had apparently as one of its purposes Adam's instruction concerning the nature of those animals over which he had been given dominion, especially those that would be in close contact with him (the "beasts of the earth," the "fish of the sea" and the "creeping things" were not included) and most likely as candidates for fellowship with man.

However, there was found in all the animal kingdom no "help meet for him" (Genesis 2:20). Adam was thus not related to the animals and had nothing in common with them. The resultant special formation of Eve out of Adam's side is the despair of theistic evolutionists, with whose peculiar approach to Biblical exegesis it simply cannot be reconciled.

Similarly, certain Bible interpreters have felt the need of what they call "pre-Adamite men," in order to reconcile the presumed evidence of man and his cultures at dates far earlier than any stretching of the Adamic genealogical lists in Genesis 5 and 11 can possibly allow. But the Bible is inflexible in this respect. Adam is the "first man" (I Corinthians 15:45,47) and Eve is the "mother of all living" (Genesis 3:20). There could be no men before the first man, nor any men living before their mother.

What, then, about Neanderthal and Pithecanthropus and the Paleolithic (Old Stone Age) and Mesolithic (Middle Stone Age) cultures? We can only conclude that, if they were truly men (as their cultural and religious artifacts often show), then they must have been descendants of Adam and, in most cases at least, descendants of Noah as well. In many cases (e.g., Ramapithecus, Australopithecus, etc.) the very fragmentary evidence is quite consistent with the view that such creatures were merely extinct species of apes.

The so-called "cave-men," such as Neanderthal man and Cro-Magnon man, were not half-brutish ape-men as many assume. They made fine cave paintings, cultivated flowers, and buried their dead. The fact that they lived in (or perhaps merely occupied) caves proves nothing regarding their stage of evolution. Many people live in caves today.

Those that were true men (Neanderthal man and Cro-Magnon man, for example, are now acknowledged by practically all anthropologists to be true *Homo Sapiens*) simply represent extinct tribes of men, no more different from other present-day tribes than many present-day tribes are from each other. They are normally found in so-called Pleistocene or Recent deposits and this fact implies that they lived after the great Flood. Chronological problems are resolved when it is realized that all dates before about 3000 B.C. are based on radiocarbon or other indirect methods, not on actual written historical records. The assumptions on which such methods are based are suspect because of their fundamental evolutionary and uniformitarian presuppositions. As discussed further in Chapter *VIII*, all such calculations are easily capable of reconciliation with the much more condensed Biblical version of human chronology.

Thus, man is a unique creation of God, entirely without evolutionary relation to the animals, and Adam was the first man, created on the sixth day of creation, much more recently than the speculative chronologies of evolutionary anthropologists have suggested. However, the present-day nations, tribes, cultures and languages of men have all been derived from the three sons of Noah, after the great Flood. "And the sons of Noah, that went forth of the ark, Shem, and Ham, and Japheth . . . These are the three sons of Noah: and of them was the whole earth overspread" (Genesis 9:18,19).

For a time after the Flood, most of Noah's descendants lived together in the Mesopotamian plain south of Mount Ararat, where the ark had landed. This, in itself, was contrary to the will of God, who had commanded them to "Be fruitful and multiply and fill the earth" (Genesis 9:1). He had also entrusted to men the responsibility of governing themselves, in stable social units, even with the prerogative of capital punishment if the crime should so warrant (Genesis 9:6).

A remarkable prophecy had also been given through Noah in Genesis 9:25–27 concerning the general characteristics which the descendants of his three sons would manifest in the post-diluvian world. Man's spiritual, intellectual and physical development was to be committed especially to Shem, Japheth, and Ham, respectively. Of Shem, he said: "Blessed be the Lord God of Shem," evidently implying that through

Shem the Lord would be blessed; the Semites would transmit the knowledge of the true God to future generations. This was fulfilled in a particular way by the nation of Israel and especially by Jesus Christ and by the Jewish believers who wrote the Scriptures.

The prophecy concerning Japheth was: "God shall enlarge Japheth, and he shall dwell in the tents of Shem." This enlargement apparently applies to both geographical expansion and intellectual growth—that is, both political and intellectual enlargement. Spiritually, however, Japheth would simply depend on Shem, dwelling in his tents. This prophecy has been fulfilled by the Aryan and Caucasian peoples, the Europeans and Americans. Again, however, it seems to have been especially fulfilled by the Greek nation, who gave science and philosophy to the world. The Greek empire was, of course, the first world kingdom of Gentile power, to be followed by the even more extensive Roman empire and by other great Japhetic nations derived from Greece and Rome. The ancient Greeks acknowledged "Iapetos" as their original ancestor and they have always been recognized as the prime example of Japhetic culture. It is also significant that, while the writers of the New Testament were all Jews, they actually wrote it in the Greek language.

Noah's other son was Ham and because of Ham's sin against his father, Noah was unable to prophesy either a spiritual (as for Shem) or intellectual (as for Japheth) contribution for his descendants. The other main aspect of man's life is physical and it is on this area that his Hamitic prophecy focuses. And again, as for Israel in the case of Shem and the Greeks in the case of Japheth, the prophecy centers especially on one particular son of Ham, Canaan, though it undoubtedly applies in general to all Ham's descendants. (The prophecy clearly was intended to be universal in scope, so those not included either as Semites or Japhethites must necessarily be considered Hamites and thus included to some degree in the prophecy concerning Ham.)

"A servant of servants shall he be to his brethren" (Genesis 9:25). This strange prophecy seems to have had a partial fulfillment in a literal sense, but a broader fulfillment in a more extended sense. The first great Hamitic nations, Sumeria and Egypt, eventually were conquered and subjugated by the Semites. The great Canaanite kingdoms, including especially the Hittite empire and later the seafaring Phoenicians, were finally defeated by Japhethites. None of these has ever risen to power since. The Canaanite tribes in Palestine were conquered by the Israelites and the African tribes by various Japhethite nations, as were the Indians of America and the Polynesians of the South Seas. The one significant exception seems to have been the Mongols, including the Chinese and Japanese (it is probable that these are Hamitic peoples, since there is no evidence that they could be either Semitic or Japhetic). Although these have never been actually subjugated by Shem or Japheth, militarily, they have until the last few years been dominated economically and scientifically by outsiders, and their own attempts at outside military enlargement always thwarted.

But there is a second, and perhaps more significant meaning to this Hamitic, or Canaanitic, "curse." Just as Shem would emphasize the spiritual side of man's nature, and Japheth the intellectual (spirit and soul, respectively), so Ham would mainly be concerned with the "body," or material aspect of man. A "servant of servants" may mean in effect a "servant extraordinary," the one who would provide physical benefits for mankind. It can be shown, for example, that Hamitic peoples were the original pioneers who settled in the world's remote regions (neither Columbus nor the Vikings discovered America—the Indians did!). Hamites invented both agriculture and animal husbandry, for man's food, and building methods, for man's shelter. They were the original shipbuilders and mariners—especially the Canaanitic Phoenicians—thus inaugurating transportation and world trade. The Phoenicians also invented the alphabet and the Chinese invented printing, for man's need for written communications. Innumerable discoveries in the fields of medicine and technology are attributable to Hamites, as well as the invention of money and banking. The list could go on and on. If man owes his spiritual heritage to Shem and his scientific and philosophical heritage to Japheth, he surely owes most of his material comforts to Ham.

There are, of course, two problems with this interpretation. Why should this vast contribution to man's material well-being be considered a "curse"? And why was it pronounced specifically relative to Canaan, instead of his father Ham—as were the prophecies relating to Shem and Japheth?

The curse in one sense, of course, applies to military and economic subjugation, as already noted. Secondly, it could be considered as relative only—that is, there was a "blessing" concerning spiritual values, but a "curse" concerning material pursuits; there was neither blessing nor curse, but merely a statement of fact concerning intellectual matters. Note, incidentally, that all of these are prophecies, that have been fulfilled, because of God's foreknowledge and also because of Noah's insight into the fundamental natures of his three sons. They do not require any extraneous action from man to assist in their fulfillment–it is a very strange type of exegesis that has led some people to attempt to justify enforced slavery of some of Ham's descendants on the basis of this passage.

Another aspect of the curse on Ham can be noted. Although he has provided so many material comforts to mankind in general, he himself has participated in few of them. The original Canaanites, the Phoenicians, Hittites, Sumerians, most of the Egyptians, Cretans, and others have either become extinct or completely lost their identities through assimilation. Their modern-day descendants, such as the Africans, the South Sea islanders, the Aborigines, the American Indians, the Copts of Egypt, the Dravidians of India, even most of the Chinese and Japanese, have lived through most of their history largely in poverty. Though Ham has been a great servant to mankind, he has mostly benefited Shem and Japheth rather than himself, just as the prophecy foretold.

The original Edenic curse, pronounced on both man and his world, stressed particularly the toil necessary to sustain himself physically in this world. The Hamitic curse likewise focused on the physical aspect of man's life and directs Ham, as it were, to take the lead in mitigating the material aspects of the Edenic curse.

Now although the prophecy undoubtedly encompassed all of Ham's descendants, it focused particularly on Canaan. This was probably because it was occasioned by the sin of Ham, who was Noah's youngest son (Genesis 9:24). Canaan was likewise Ham's youngest son (Genesis 10:6). Noah thus emphasized that just as Ham's sin had dishonored and grieved his father, so Ham likewise would suffer through his own sons, the curse extending even to his youngest son. The reference to Canaan serving his brethren also served to show that the prophecy applied not just to Noah's own sons, but also to their descendants.

It should be stressed of course, that this prophetic tri-furcation of mankind has a general application to the nations as a whole, not to all individuals in the nation. An individual Japhethite, for example, might be adept in material technology and an individual Hamite very spiritual. The prophecy is general, not specific.

However, before the prophecy could begin to be fulfilled, and also before God's purpose for man on the earth could be accomplished, men must be forced to separate and to develop their own cultures and civilizations. Although communication and cooperation among men would be necessary for these purposes, it soon appeared that too easy communication would lead to the wrong kind of cooperation, not to control man under God, but rather to unite man in rebellion against God.

This rebellion was also led by Hamites, especially by Nimrod, at the city of Babel. The events are described in Genesis 11:1–9 and the leadership of Nimrod in Genesis 10:8–12. It is probable that there were approximately 70 family clans at Babel, as indicated by the 70 original tribes listed in Genesis 10:1–32. These included Semites and Japhethites, as well as Hamites.

Rather than moving out to "fill the earth," as God had commanded, they all prepared to unite together in the form of a great city, in which they could "make us a name" (Genesis 11:4). This decision involved a direct rebellion against God's command and thus indicated that at least their leaders, especially Nimrod, no longer feared God.

They seemingly had been convinced, perhaps by some form of occult communication with demonic spirits, that they could successfully follow Satan in his continuing rebellion and thus be free from God's restraint and especially His Hamitic curse. They built the first great temple "tower, whose top . . . unto heaven" (Genesis 11:4) as the central headquarters of this rebellion. They probably emblazoned the various astrological emblems in the shrine at the top, thus dedicating the tower "unto heaven"—that is, to the worship of the host of heaven. The stars were identified with the host of Satanic angels. This primeval astrological system is the fountainhead of all occultism, polytheism, and false religion in general. Babylon was the "mother of harlots" (Revelation 17:5).

In any case, it was from Babel that God forcibly dispersed the rebels, by a mighty miracle. "The Lord did there confound the language of all the earth: and from thence did the Lord scatter them abroad upon the face of all the earth" (Genesis 11:9).

The skeptic may question this record, but let him devise a better explanation for the origin of human languages! The languages of even the most "primitive" tribes are extremely complex and are removed by a great gulf from the chatterings of the most "advanced" apes, as well as other animals. There is neither any evidence nor any explanation for any assumed naturalistic evolution of human language.

The attribute of language—the ability to articulate and communicate even abstract concepts—is the most basic aspect of a human culture. Physiological distinctives such as skin color are of minor importance compared to language as the cause of divisions among the nations and tribes of mankind. When God decided to enforce a separation and scattering of men, He did so by the most effective way possible, confusing their tongues. After they were separated into distinct tribal units, then it was possible for distinctive physiological characteristics, hitherto inhibited by the intermarrying at Babel, to become fixed genetically, by tribal inbreeding. Thus the different physical characteristics of different national groups were indirectly the result of the confusion of tongues.

The seventy original nations, as listed in Genesis 10, have proliferated into over 2000 tribes and languages. Since this primeval dispersion there have been many attempts by strong men to unite all nations under their rule, as well as efforts by politicians to establish voluntary unions of all nations, but every one has failed. This is because "the Most High divided to the nations their inheritance, when He separated the sons of Adam" (Deuteronomy 32:8). God has "made of one blood all nations of men for to dwell on all the face of the earth, and hath determined the times before appointed, and the bounds of their habitation" (Acts 17:26).

It is well to recognize that, while the Scriptures place high importance on the distinctiveness of nations and tribes as such, they never once mention the concept of *race*. Biblically there is only one race—the human race. There is only one *kind* of man—that is, *mankind*. "God hath made of one blood all nations of men."

The idea of race is strictly a category of evolutionary biology, not of Scripture at all. The threefold division of mankind into Japhethites, Semites, and Hamites is not a *racial* division, but rather of three different streams of nations. The biological entity known as "race," on the other hand, is supposed to be a sub-species in the process of evolving into a new species, with a long evolutionary history of its own. Modern racism has always found its strongest and most vicious expression among doctrinaire evolutionists—men such as Karl Marx, Adolph Hitler, and other such advocates of group struggle and survival of the fittest.

A real understanding of man in relation to his world will never be attained, nor solutions to his problems ever achieved, as long as our educational and political leaders persist in thinking of them in evolutionary categories. Man is not an evolved animal and his cultures and institutions have not been developed from the herd-instincts of animals. Rather, he is a unique creation, made in the image of God, and his tribes and nations represent divisions established by God for man's own good and for the ultimate accomplishment of His divine will on the earth, not through their own devices but by His power and grace.

CHAPTER VI
THE PUZZLING ROLE OF THE STARS ABOVE

The invention of the telescope is a very recent event in earth history, but it has made it possible for man to discover that the amazing statements of the Bible about the universe were true after all. For example, the Scriptures teach that the cosmos is infinite in size and the stars are innumerable, at least in so far as man can measure. God said, in Isaiah 55:9: "For as the heavens are higher than the earth, so are my ways higher than your ways, and my thoughts than your thoughts." Similarly, David said: "For as the heaven is high above the earth, so great is His mercy toward them that fear Him" (Psalm 103:11). The only comparisons appropriate for the infinitude of space are the omniscience and infinite grace of God.

The study of the heavens ought by all means to convince men of the necessity of a Creator. As the great astronomer Herschel used to say: "The undevout astronomer is mad!" Yet today it is sadly true that there are perhaps fewer creationists among professional astronomers than in almost any other branch of science. Theories of stellar and galactic evolution have become so inextricably interwoven with the study of the stars that it is almost impossible to separate between fact and speculation in modern-day astronomical writings.

For some time now, astronomical theorists have been divided into two camps—those favoring Gamow's "big-bang" theory and those supporting Hoyle's "continuous-creation" theory. Both theories are evolutionary theories demanding an immensity of time in which to function. Neither is capable of scientific proof since both deal with non-reproducible events of past history. How does one study *experimentally* the evolution of a universe?

In addition, there are many subsidiary theories relating to the evolution of particular types of stars and galaxies, which theories are themselves more or less independent of the basic theory. These are also not susceptible of experimental proof. Cosmogony seems to be a sort of game that astronomers play, a tongue-in-cheek charade in which only the initiates know the rules and the spectators stand in awe.

It will help the outsider to maintain his balance in the sea of astro-evolutionary speculation if he will keep reminding himself that no astronomer has ever *seen* one kind of star evolve into another kind of star or one type of galaxy into another. As long as men have been observing the stars they have *never* seen them evolving! The starry heavens look exactly the same today as they did when Galileo and even Nimrod first looked at them—occasional "novas" and dying comets and meteorites excepted. Ideas about their "evolution" must always remain speculation—and nothing else.

It is strange that astronomers are so reluctant to consider the possibility that the heavens may have been specially created in their present form. After all, science is supposed to be based on observation, what one actually *sees* taking place in the real world. Astronomers never *see* evolution taking place. What they see is the essential stability of the starry heavens. Since this is exactly what would be expected on the basis of an initial completed creation, why should they not conclude that creation is the most reasonable explanation for what they see?

It seems the only answer to this conundrum is again to be found in the words of Romans 1:28: "They did not like to retain God in their knowledge." Nevertheless it remains true that there is not a single *fact* of observational astronomy which cannot be satisfactorily understood in terms of the special creation of the stars in their present forms in the beginning. There are many such facts, on the other hand, which cannot be satisfactorily explained in terms of any of the current evolutionary theories of the origin of the universe. That being the case, we conclude that the special creation theory is the best theory, strictly on the scientific merits of the case.

The problem is completely settled, of course, by the Scriptures. The fact that some men do not believe the Bible is irrelevant. The Bible is still true, regardless of what men may think! And the Bible says: "By the word of the Lord were the heavens made, and all the host of them by the breath of His mouth . . .

For He spake, and it was done; He commanded, and it stood fast" (Psalm 33:6,9). "He commanded, and they were created" (Psalm 148:5).

The stars, in fact, were made on the fourth day of creation week, along with the sun and moon. "And God made two great lights, the greater light to rule the day, and the lesser light to rule the night: He made the stars also . . . And the evening and the morning were the fourth day" (Genesis 1:16,19).

The formation of these bodies to "give light upon the earth" (Genesis 1:17) was, of course, subsequent to the creation of light itself. On the first day of creation, God had said: "Let there be light: and there was light" (Genesis 1:3). This light was visible light, since it was to "divide the light from the darkness." However, it obviously did not emanate from the sun, moon, and stars, since these were not made until the fourth day. There is no way now of determining the nature of this initial source of visible light, since it was later delegated, as it were, to the heavenly bodies. It may well have emanated from the theophanic presence of God Himself.

As far as effects were concerned, the primeval light of the first day and the "light bearers" of the fourth and all succeeding days produced the same results. They "divided the light from the darkness" (Genesis 1:4,18). This division has been accomplished since the fourth day by the rotation of the earth on its axis; presumably the same was therefore true on the first three days.

Thus, two basic movements began concurrently on this first day. One was the diurnal rotation of the earth, which is the most important cyclic movement in the universe, as far as man is concerned. The other was the motion of light waves, traveling at the ultimate of all speeds, the velocity of light.

Motion is the manifestation of energy. It is significant that these two fundamental types of motion followed directly when "the Spirit of God *moved* on the face of the waters" (Genesis 1:2). The primeval earth was created formless and empty, covered with water and surrounded by darkness. But then God's Spirit proceeded to "move" (or, more literally, "vibrated") over the face of the waters. The result was the outraying of vibrating light waves and the spinning of the earth on its axis.

Although the emphasis in Genesis is on visible light, it is probable that the entire spectrum of electro-magnetic energy was initiated at the same time. Certainly by the time of the fourth day this must have been true, since the sun, moon, and stars were made then, and all types of electro-magnetic phenomena are associated with these bodies. The continuity of other forms of electro-magnetic energy with visible light in the spectrum would also indicate that the entire spectrum was included in God's initial command for the light to appear.

There are problems, of course, involved in the teaching of Genesis 1:16 that the stars were made only on the fourth day of creation. The first is Genesis 1:1: "In the beginning God created the heaven and the earth." This statement obviously applies to the first day, not the fourth.

However, the stars are not "heaven," but rather the "host of heaven" (Genesis 2:1, Deuteronomy 4:19). They were placed as "lights in the firmament of the heaven," (Genesis 1:15) but they were not heaven itself. The term "heaven" (Hebrew *shemayim*) thus is essentially equivalent to "space." It can be used either of space in the general sense or of a particular space, such as the space where the stars were placed or the space where the birds fly.

The term "firmament" is similar (Hebrew *raqia*). It means, literally, "thinness" or "that which is stretched out." The phrase "firmament of the heaven," therefore, connotes "stretched-out space."

Another problem that has been associated with Genesis 1:16 is found in Job 38:4–7, where God asks: "Where wast thou when I laid the foundations of the earth . . . When the morning stars sang together, and all the sons of God shouted for joy?" This passage has been interpreted as teaching that the stars were already in existence when the foundations of the earth were laid and therefore directly contradicts Genesis 1:16, which says the stars were made on the fourth day.

However, it should be clear that figurative language is used here, since "morning stars" do not literally "sing together." In fact, the Hebrew construction makes it clear that the "morning stars" are actually the same as the "sons of God." The latter were the created beings called angels and they are frequently associated in Scripture with the stars. Both the stars and the angels are, in fact, called "the host of heaven" in many places in the Bible. Whatever the reason for this symbolic association may be, it is obvious that Job 38:7 refers to the angels, and not to actual stellar bodies.

A third problem is scientific in nature. If the stars were made on the fourth day, and if the days of creation were literal days, then the stars must be only several thousand years old. How, then, can many of the stars be millions or billions of light-years distant, since it would take correspondingly millions or billions of years for their light to reach the earth?

This problem seems formidable at first, but is easily resolved when the implications of God's creative acts are understood. The very purpose of creation centered in man. Even the angels themselves were created to be "ministering spirits, sent forth to minister for them who shall be heirs of salvation" (Hebrews 1:14). Man was not some kind of afterthought on God's part at all, but was absolutely central in all His plans.

The sun, moon, and stars were formed specifically to "be for signs, and for seasons, and for days, and years," and "to give light upon the earth" (Genesis 1:14,15). In order to accomplish these purposes, they would obviously have to be visible on earth. But this requirement is a very little thing to a Creator! Why is it less difficult to create a star than to create the emanations from that star? In fact, had not God created "light" on Day One prior to His construction of "lights" on Day Four? It is even possible that the "light" bathing the earth on the first three days was created in space as en route from the innumerable "light bearers" which were yet to be constituted on the fourth day.

The reason such concepts appear at first strange and unbelievable is that our minds are so conditioned to think in uniformitarian terms that we cannot easily grasp the meaning of creation. Actually, real creation necessarily involves creation of "apparent age." Whatever is truly created–that is, called instantly into existence out of nothing–must certainly look as though it had been there prior to its creation. Thus it has an appearance of age.

This factor of created maturity obviously applies in the case of Adam and Eve, as well as of the individual plants and animals. There is nothing at all unreasonable in assuming that it likewise applies to the entire created universe! In fact, in view of God's power and purposes, it is by far the most reasonable, most efficient, and most gracious way He *could* have done it.

We still do not know the full answer to the problem of the total purpose of all the stars. Especially is this true of the innumerable stars that can only be seen through telescopes. If astronomic distances are as great as astronomers believe, then it is quite impossible for man himself ever to reach the stars by space travel. Even the closest star is four light-years distant, over 10,000 times as far away as the sun. The stars that are visible to the naked eye are, of course, valuable for navigation as well as beauty, but these only constitute an infinitesimal fraction of the total number of stars. What, then, was the purpose God had in creating all the others?

Certain tentative and partial answers to this question can be suggested. In the first place their amazing number and variety bear an ever-growing witness to the infinitude of God's power and ingenuity. "The heavens declare the glory of God, and the firmament sheweth His handiwork" (Psalm 19:1).

Secondly, since in Scripture stars are frequently associated with angels, it may be that the stars are in some way involved in the ministries of the angels. It is interesting that the Bible speaks of an "innumerable company of angels" (Hebrews 12:22), just as it similarly speaks of the stars in the declaration: "the host of heaven cannot be numbered" (Jeremiah 33:22). This possible association of angels with the stars, incidentally, is the *only* suggestion that Scripture makes concerning intelligent life on other worlds. There are definitely no men, or man-like intelligences, living on other planets or stars, in so far as the Bible indicates. "The heaven, even the heavens, are the Lord's: but the earth hath he given to the children of men" (Psalm 115:16).

Lastly, it may be that, even though the stars are beyond man's reach in this present age, they will be accessible to him in "the ages to come" (Ephesians 2:7). The stars, as well as the earth, the sun, and the moon, were created to last forever (note Psalm 104:5; Psalm 148:3–6; Jeremiah 31:35,36; etc.). Those who share in the resurrection of the just will have resurrection-bodies which, though truly real and physical, will yet be "as the angels of God in heaven" (Matthew 22:30). They will be immortal and incorruptible bodies, not subject to death and deterioration, as are our present bodies (I Corinthians 15:51–53). Similarly, just as the angels, they will no longer be subject to either the gravitational or electro-magnetic forces of the cosmos. When the Lord comes, we "shall be caught up together" (I Thessalonians 4:17), "to meet the Lord in the air." We shall have a body "like unto His glorious body" (Philippians 3:21) and, thus, like Him, will be able easily to "ascend up far above all heavens" (Ephesians 4:10). Not being subjected to the forces of the physical world, our movements will not be controlled by them, and thus our "spiritual bodies" (I Corinthians 15:44) can move with velocities far exceeding the speed of light. Consequently, inter-galactic travel will be perfectly feasible for redeemed men in future ages, just as it is for the angels even in this present age. It may well be a fitting activity and responsibility for men in the *eternal* future to explore and develop the *infinite* universe that came from the creative hand of the omniscient God.

The unending variety and intricacy of the galaxies of the heavens are, even now, a marvelous discovery made through the gigantic optical and radio telescopes of the present day. "There are also celestial bodies, and bodies terrestrial, but the glory of the celestial is one, and the glory of the terrestrial is another. There is one glory of the sun, and another glory of the moon, and another glory of the stars; for one star differeth from another star in glory" (I Corinthians 15:40,41).

We need now to return briefly to a consideration of possible changes that have taken place in the stars since they were created. At the end of the six days of creation, "God saw everything that He had made, and behold, it was very good" (Genesis 1:31). That this statement included the heavens as well as the earth is evident from both Genesis 2:1 and Exodus 20:11. The planets, however, as well as asteroids and meteorites, give evidence of having experienced an abundance of violent volcanic and seismic activity since they were created, and the occasional appearance of novas and supernovas among the stars indicates that these also are subject to destructive forces.

On the earth such phenomena are attributable basically to God's Curse on man's dominion because of his sin. To what extent this Curse affected the rest of the universe is not explicitly revealed, except that the Scripture does say that "the whole creation groaneth and travaileth in pain together until now" (Romans 8:22).

There is also, of course, the possibility that the sin of Satan, who evidently once had dominion over the stars of heaven, led to a curse on his dominion just as Adam's sin led to the Curse on the earth. He had been the "anointed cherub that covereth" (Ezekiel 28:14), evidently the highest among the cherubim, who are those in the angelic hierarchy closest to the throne of God Himself (Psalm 99:1). But he had said: "I will ascend into heaven, I will exalt my throne above the stars of God" (Isaiah 14:13). Furthermore, he had led with him in his rebellion "the third part of the stars of heaven" (Revelation 12:4). God, therefore, said "I will cast thee to the earth" (Ezekiel 28:17), and ultimately "down to hell, to the sides of the pit" (Isaiah 14:15).

Satan thenceforth became "the prince of the power of the air" (Ephesians 2:2) and has concentrated his efforts and those of his angelic followers on trying to defeat God's plans for man, especially by enlisting man (with his unique powers of procreating his kind) in his own battle against God.

God's holy angels, on the other hand, have continued faithful to God and are continually "doing his commandments" (Psalm 103:20) and especially keeping watch over the "heirs of salvation." It is clear, therefore, that there is a continuing cosmic warfare between "Michael and his angels" and "the dragon and his angels" (Revelation 12:7). Aspects of this conflict are glimpsed occasionally in Scripture (Daniel 10:5, 12–13,20; Psalm 34:7; Jude 9; etc.).

The physical stars, which are somehow associated with the spiritual host of heaven, may thus be also involved in this heavenly warfare. The "stars" associated with the solar system, such as the planets and asteroids (and it should be remembered that the term "star" in Biblical usage applies to any heavenly body other than the sun and moon) would be particularly likely to be involved, in view of the heavy concentration of angels, both good and evil, around the planet Earth.

There are a number of Biblical references indicating that in some way the stars may actually participate in human battles (Numbers 24:17; Judges 5:20; Revelation 6:13; 8:10; etc.). Such passages may all be simply figurative, but then again they may not. In any case, the possibility is at least open that the fractures and scars on the moon and Mars, the shattered remnants of an erstwhile planet that became the asteroids, the peculiar rings of Saturn, the meteorite swarms, and other such features that somehow seem alien to a "very good" universe as God must have created it may have been acquired later. Perhaps they reflect some kind of heavenly catastrophe associated either with Satan's primeval rebellion or his continuing battle against Michael and his angels.

Immanuel Velikovsky and other modern writers have stressed the possible significance of the host of ancient traditions and myths dealing with the "wars of the gods." Such "wars" and "gods" were always for some reason associated with the various stars, whose names they shared. The long fascination of men of all nations with pagan astrology can only be understood if it is recognized that there is some substratum of truth in the otherwise strange notion that objects billions of miles away could have any influence on earthly events. Certainly the physical stars as such can have no effect on the earth, but the evil spirits connected with them are not so limited. Furthermore, the well-documented association of certain "U.F.O." sightings with occultic influences and tendencies suggests that the "rulers of the darkness of this world" (Ephesians 6:12) are increasingly imaginative in their battles for the minds of men.

Angels, both good and bad, can be shown Biblically to have considerable knowledge and power over natural processes and, thus can in many cases either cause or prevent physical catastrophes on earth and in the heavens. In any event, this type of cause warrants further research as a potential explanation for apparent disturbances in the stars and planets since their creation.

In concluding this chapter, we come finally to consider that first and most basic of all Scriptures, Genesis 1:1: "In the beginning, God created the heaven and the earth." We have already noted that the word "heaven" meant, essentially, "space." Similarly, it is evident that "earth" at this point meant, essentially, "matter." As we have seen, there were then no other physical bodies yet in existence. The earth was the only "matter" created. As yet, however, it had neither structure nor occupant, living or non-living. It was "without form and void" (Genesis 1:2). The remainder of Genesis One tells how this initially formless and empty matter was structured and equipped to accomplish its purpose and function as man's home.

To all effects, therefore, Genesis 1:1 can be understood as saying: "When time began, God created space and matter." The space-matter-time continuum, which constitutes the basic framework of our universe, was thus brought into existence as God's first recorded act of special creation. "Time" thus proceeded to flow onward forever. "Space" was stretched out toward infinity. And everywhere throughout space and time, "Matter" began to appear, first as the formless earth, then as a matrix of water sustaining the basic elements of matter in the earth, then as light energy piercing the darkness, then as an atmosphere separating two great hydrospheres, then as solid materials emerging from the watery matrix, such materials perhaps resting for the first time on what the Bible calls "the foundations of the earth" deep in the earth's core, next as complex replicating materials all over the earth's surface forming its covering of vegetation, and then, finally, as an infinite array of heavenly bodies dotted everywhere throughout the universe, in an equally infinite variety of sizes, shapes, groupings, compositions and activities.

As one scans the starry heavens, he can hardly refrain from asking with the Psalmist: "When I consider thy heavens, the work of thy fingers, the moon and the stars, which thou hast ordained; What is man, that thou art mindful of him?" (Psalm 8:3,4). And yet, in all the created physical universe, man alone has "the image of God." Standing halfway in size between the incredibly wonderful world of sub-atomic space and the gigantic and innumerable stars of cosmic space, man's structure is the most complex and intricate system ever created. Of man, as represented and redeemed by the Son of Man, it is witnessed: "Thou madest him to have dominion over the works of thy hands; thou hast put all things under his feet" (Psalm 8:6).

CHAPTER VII
THE STRANGE DELUSION OF EVOLUTION

We have now considered a number of aspects of the problem of origins—the origin of the universe, the origin of the solar system, the origin of life, the origin of the various kinds of organisms, the origin of man, and the origin of man's basic languages. In every case, the Genesis record is clear in its teaching that these are all special creations of God; none came by any process of evolution.

We have also seen that the innumerable evidences of order and complexity in the world point clearly to a great Designer, and that the fundamental nature of all physical processes likewise points to a primeval period of special creation. The clear testimony of true science thus supports the explicit testimony of Genesis that the world and all things therein came into existence by special creation, not by innate processes of development.

However, many religious people today have felt they should somehow accommodate evolution in their system of thought. The wide acceptance of evolution in the academic and political worlds has made it uncomfortable for them to take a stand against it, especially since such a stand requires some study and effort to be effective and may also involve willingness to be ridiculed and pressured by their more "liberal" evolutionist colleagues. Accordingly, many Christians have been inclined to go along with some form of theistic evolution and others have thought they could, like ostriches, ignore the whole problem. They find it more convenient to assume that Scripture can somehow be reinterpreted to harmonize with evolutionary philosophy.

For those, however, for whom Biblical and theological considerations are of first importance, there can be no compromise with the evolutionary system. We need to recognize that science as such can give no conclusive determination as to origins, even though the *facts* of science do lend themselves to interpretation better in terms of creation than of evolution. But the Christian is not limited to empirical science on the question; he has the advantage of authoritative revelation from the Creator Himself, in the Holy Scriptures.

For Christians, therefore, considerations such as the following should settle the question, regardless of any remaining unsolved scientific questions. Listed below are seven basic Biblical and theological objections to any Biblical compromise with evolution. So far as we know, no "Christian evolutionist" has yet attempted a serious answer to even one of these difficulties.

(1) *Evolution contradicts the Bible record of a finished creation.*

The fundamental premise of evolutionary philosophy is that the origin and development of all things can be understood in terms of basic natural laws and processes which can be studied in operation right now. This assumption flatly contradicts the Biblical statement that "God rested from all His work which He created and made" (Genesis 2:3) after the six days of creation. "In six days the Lord made heaven and earth, and on the seventh day He rested, and was refreshed" (Exodus 31:17). "The works were finished from the foundation of the world" (Hebrews 4:3). Scientifically, this statement of the completion of the creative process anticipated by thousands of years the discovery of the law of conservation of mass-energy.

(2) *Evolution contradicts the doctrine of fixed and distinct kinds.*

If evolution is really true, then all kinds of plants and animals have developed naturally from a common ancestor. Consequently there is no real permanence of kinds, but a perpetual state of flux and an organic continuity of all forms of life. However, the Bible says "All flesh is not the same flesh: but there is one kind of flesh of man, another flesh of beasts, another of fishes, and another of birds" (I Corinthians 15:39). God, in His omniscience and omnipotence, was able to make things the way He wanted them to be, and He intended them to stay that way. "God made the beast of the earth after his kind, and cattle after their kind, and every thing that creepeth upon the earth after his kind: and God saw that it was good" (Genesis 1:25). "God giveth it a body as it hath pleased

Him, and to every seed his own body" (I Corinthians 15:38). This primevally-ordained fixity of kinds is illustrated in the law of biogenesis—life comes only from life and like begets like.

(3) *Evolution is inconsistent with God's omniscience.*

The supposed history of evolution is filled with trial and error, misfits, evolutionary blind alleys, and meaningless extinctions. The supposed raw materials of evolution are provided by random mutations, not by ordered planning. If the goal of the evolutionary process was man, why did God take so long to get to the business at hand? What was the need for dinosaurs to rule the earth for 100 million years only to become extinct 70 million years before man evolved? Such an aeon-long process of random evolutionary meandering is the most wasteful and inefficient method that one could possibly devise by which to create a world. The omniscient Creator would certainly not be so foolish as this. "God is not the author of confusion" (I Corinthians 14:33). God has commanded "Let all things be done decently and in order" (I Corinthians 14:40), and it would be strange for Him to set such an incredibly poor example in His own work.

(4) *Evolution is contrary to God's nature of love and mercy.*

Unnumbered billions of living creatures must have died in the history of the evolutionary process, if evolution is really true. The fossil record is filled with evidence of injury and disease, of suffering and violent death, on a gigantic scale. How could the "God of all grace" (I Peter 5:10) possibly be responsible for such a cruel spectacle as this? "One sparrow shall not fall on the ground without your Father" (Matthew 10:29). It would be infinitely more gracious and merciful, as well as more efficient, for God to create all things complete and fully developed right from the start, exactly as the Bible says He did.

(5) *Evolution contradicts the universal principle of decay.*

Ever since God said "Cursed is the ground" (Genesis 3:17), the "creation itself" has been waiting to "be delivered from the bondage of corruption" (Romans 8:21). "All flesh is grass, . . . the grass withereth, the flower fadeth" (Isaiah 40:6-7). "The earth shall wax old like a garment, and they that dwell therein shall die in like manner" (Isaiah 51:6). There is in effect a universal principle of disintegration and death, both the physical creation ("earth shall wax old") and the living world ("all flesh is grass"). This is nothing less than the Curse, pronounced by God on man's entire dominion because of man's sin, reflected in the scientific realm by the universal law of increasing entropy. It is obvious that the concept of a universal process of order increasing from molecule to man is incompatible with a universal process of decay and a decreasing order.

(6) *Evolution is incompatible with Christian ethics.*

The essence of the evolutionary process is survival, because obviously no organism can contribute to evolution unless it survives and reproduces. The concept of natural selection entails a struggle for existence and survival of the fittest. The weak and misfits are exterminated, the strong and fertile survive. If God had anything to do with the evolutionary process, it does seem strange that He would utilize a method which squarely contradicts the system of ethics He established for the man He created by this process. Jesus said "Ye have heard that it hath been said, "An eye for an eye, and a tooth for a tooth: But I say unto you, That ye resist not evil: but whosoever shall smite thee on thy right cheek, turn to him the other also" (Matthew 5:38,39). The chief good of evolution is struggle and survival, but the essence of Christianity is sacrifice and death.

(7) *Evolution produces anti-Christian results.*

Jesus said: "A good tree cannot bring forth evil fruit, neither can a corrupt tree bring forth good fruit" (Matthew 7:18). Evolution is the root of atheism, of communism, nazism, behaviorism, racism, economic imperialism, militarism, libertinism, anarchism, and all manner of anti-Christian systems of belief and practice. A solid faith in a personal, sovereign Creator, on the other hand, leads to a strong sense of responsibility before God and therefore eventually to an awareness of one's need for a personal Saviour.

Many other Biblical and theological difficulties with the theory of evolution could be listed if necessary. The foregoing summary, however, should make it clear that theistic evolution cannot be harmonized with Biblical Christianity.

Brief notice should be taken, however, of a related concept. A sizable number of Christian intellectuals, aware of the unsavory connotations of theistic evolution to conservative Christians, have substituted for it the semantic curiosity called *progressive creation.*

To some writers, of course, this term is to all intents synonymous with theistic evolution. They prefer it merely because it sounds better, especially to alumni and boards of trustees who have become alarmed over reports of evolutionary teaching in their Christian colleges and church literature.

More precisely, however, progressive creation is a concept which allows isolated acts of creation scattered throughout the geologic ages. For the most part the standard system of long evolutionary ages is retained intact, with the modification that the otherwise uncrossable gaps in the fossil record are assumed to be bridged by creative interjections as needed. Thus the progressive creationists do give a sort of "nod to God" now and then, and they consider this an adequate accommodation of science to Scripture.

However, the same theological fallacies of theistic evolution as listed above also apply to progressive creation, though not always in the same degree. There is still the same spectacle of suffering and death, the same trial-and-error meandering, the same self-centered survival ethic, the same basic system with all its corrupt fruit.

As a matter of fact, in one important sense progressive creationism is actually more dishonoring to God than is the idea of theistic evolution. That is, theistic evolution at least assumes that God was able to plan and control the entire evolutionary process from the beginning. Progressive creation, on the other hand, implies that God's power was inadequate to plan and energize the whole program at one time. It was necessary every few million years or so for Him to come down and give new direction and a fresh injection of creative energy to the normal evolutionary activity. Thus, progressive creation, though presenting a better image than theistic evolution in its terminology, is even more objectionable to true creationists in its theology.

It seems completely impossible, for that matter, for the God of the Bible to co-exist with the geological age system at all, regardless of whether they are identified with a theistic evolutionary or a progressive creation framework. The geologic ages are identified and dated by the fossils contained in the sedimentary rocks. The fossil record also provides the chief evidence for the theory of evolution, which is in turn the basic philosophy upon which the sequence of geologic ages has been erected. The evolution-fossil-geologic-age system is thus a closed circle which comprises one interlocking package. Each goes with the other two.

More clearly than anything else, however, the fossil record, which is the sole support of both the geologic-age system and the theory of evolution, speaks of *death.* More often than not, it speaks of sudden and violent death, as witnessed by the great fossil graveyards found all over the world and in all geological "ages."

The Bible-believing Christian must realize that, if he accepts the geologic-ages system, he is implicitly accepting the whole evolutionary package which is synonymous with it. He is accepting the billion-year reign of suffering and death in the world, including the death of even extinct tribes of men such as Homo Erectus and Neanderthal Man, who lived and died long before Adam, if the evolutionary chronology is right.

Yet the Scriptures teach plainly that there was no suffering and death in the world before Adam sinned. "Wherefore, as by one man, sin entered into the world, and death by sin" (Romans 5:12). Death came into the world only when sin came into the world—not long ages before. "By man came death" (I Corinthians 15:21).

Consequently, the vast fossil record, comprising as it does, a worldwide cemetery preserved in stone for men everywhere to see, is not at all a record of the gradual evolution of life, but rather of the sudden destruction of life. It must have been formed sometime after man first sinned, and therefore must have been formed cataclysmically, not gradually. Its testimony is not as a witness to an evolving world, but rather as a witness that "the world that then was, being overflowed with water, perished" (II Peter 3:6).

The fossil-bearing geologic column, as well as the universal law of increasing entropy, both testify of God's judgment on sin. God, as sovereign Creator and Sustainer of the universe, is not simply an absentee landlord as some have suggested. He is "not far from every one of us: For in Him we live and move and have our being" (Acts 17:27,28).

God, being God, had a high purpose in the creation of the world and in the creation of the eternal spirit in each individual person. And, being God, He cannot fail in that purpose.

The age-long reign of sin and death can, therefore, only be a temporary intruder in His universe. God has allowed "sin for a season" not only because man, in His own image, must be responsible before God, but also in order that man might come to know the Lord as Creator and as Redeemer.

Man is not an upward-evolving animal but, rather, a lost sinner under the condemnation of death. Being bound himself in the universal "bondage of corruption," there is nothing any individual can do to deliver and save either himself or others. Even such an intellectual and moral giant as Paul had to say: "I find then a law, that, when I would do good, evil is present with me. For I delight in the law of God after the inward man: But I see another law in my members, warring against the law of my mind, and bringing me into captivity to the law of sin which is in my members. O wretched man that I am! who shall deliver me from the body of this death?" (Romans 7:21-24).

It is at such a point that God reveals Himself as a God of grace and cleansing, as well as a God of power and judgment. "For as in Adam all die, even so in Christ shall all be made alive" (I Corinthians

15:22). "Where sin abounded, grace did much more abound: That as sin hath reigned unto death, even so might grace reign through righteousness unto eternal life by Jesus Christ our Lord" (Romans 5:20,21). Through the conquering grace of the Redeeming Creator, each believing sinner is forgiven and saved forever through faith in Christ, "who was delivered for our offenses, and was raised again for our justification" (Romans 4:25).

Then ultimately, when Christ comes again to bring to consummation His great purpose in creation and redemption, will be fulfilled the promise: "The creation also shall be delivered from the bondage of corruption into the glorious liberty of the children of God" (Romans 8:21).

CHAPTER VIII
THE WONDERFUL DAY WHEN THE WORLD WAS BORN

The question of the duration of geologic time is undoubtedly the most vexing problem confronting the Biblical creationist. Most geologists insist that the earth is about five billion years old, that life evolved probably three billion years ago, and even human life at least a million years ago. Yet the Bible seems clearly to teach that all things were created only about six thousand years ago.

From six thousand to five billion—this is how much the earth has "aged" in little more than a century! If the Bible is really wrong on this, it amounts to almost a million-fold mistake! And if it is mistaken this much in its very foundation—the chronologic framework of history—then how can we rely on it anywhere else? Writers who are unable to record sober facts of history correctly are not likely to inspire confidence when they forecast events of the eternal future.

It is significant, however, that the Bible's historical data, including its chronology, have been abundantly verified by the archaeological research of the past century, at least to as far back as the time of Abraham, about 2000 B.C. Although there are differences of opinion between schools of archaeological thought as to a precise Palestinian chronological scale, there is no question any more as to the general accuracy of Biblical history from Chapter 12 of Genesis forward.

That being the case, it seems strange that most modern scholars have rejected the historicity of Genesis 1–11. These first eleven chapters of Genesis appear to be written as sober, continuous history, and Genesis 11 merges directly into Genesis 12, at the time of Terah and Abram. Why, therefore, should they not likewise be accepted as historical?

The later Biblical writers frequently referred to the events of Genesis 1–11 as true history. The Genesis record of creation was verified by God Himself, as He gave the ten commandments (Exodus 20:8–11; 31:16–17). The genealogical records of Genesis 5 and 11 were accepted and repeated by the author of I Chronicles in his first chapter. Job 31:33 refers to the sin of Adam, and Isaiah 54:9 to the waters of the Flood. Ezekiel 14:14,20 mentions the righteousness of Noah, and Moses recalls the separation of the nations at Babel, in Deuteronomy 32:8. The various primeval nations of Genesis 10 are encountered frequently in various parts of the Old Testament. The creation is a prominent theme of many of the Psalms, and is often referred to in other books of the Old Testament.

But it is in the New Testament that these first eleven chapters of Genesis are most frequently mentioned. There are no less than 80 quotations or clear allusions to these chapters found in the New Testament writings. Furthermore, every one of the eight different writers of the New Testament (nine, if someone other than Paul wrote Hebrews) refers to at least one event or person in Genesis 1–11. Each chapter of these eleven is mentioned at least once in the New Testament. Perhaps most significant of all is the fact that Jesus Christ Himself referred to Genesis 1 (Matthew 19:4), Genesis 2 (Mark 10:7–9), Genesis 3 (John 8:44), Genesis 4 (Luke 11:51), Genesis 6 (Matthew 24:37,38), and Genesis 7 (Luke 17:27). Note the list in Appendix B.

Thus it is obvious that one can logically reject the historicity of Genesis 1–11 only if he likewise rejects the inerrancy of the rest of the Bible as well, and even the infallibility of Christ Himself. Many modern-day religious liberals and even some supposedly conservative Christians have done exactly that. Most Christians, however, are unwilling to go this far. Some try to avoid the issue altogether, but this tactic almost inevitably is a prelude to compromise.

The only Bible-honoring conclusion is, of course, that Genesis 1–11 is actual historical truth, regardless of any scientific or chronologic problems thereby entailed. There may be certain unsettled problems in the chronologies—for example, one or more gaps may exist in the patriarchal lists of Genesis 5 and 11 or the ages of the patriarchs themselves may have been miscopied by the ancient scribes in some instances—but any such questions involve very minor adjustments at the most. We must conclude that the Bible teaches unequivocally that Adam was created several thousand years ago, and no more. There is no legitimate exegetical justification for any accommodation with the supposed million-year duration of man's existence on the earth. To stretch the 2000 years and 20 patriarchs from Adam to Abraham, as indicated in Genesis 5 and 11, into the million or more years and 30,000 generations demanded by evolutionists since the first true man is to caricature these chapters of Genesis and to render meaningless any objective criteria of Bible exegesis.

As far as pre-human history is concerned, the intellectual battleground is of course the first chapter of Genesis. The "naive literal" reading of this chapter (to use the supercilious terminology of Bernard Ramm, in his influential book *The Christian View of Science and Scripture*) would indicate of course that all things were made in six days. Geologists, however, insist on five billion years, and "this is an hard saying"!

Is there any way by which the six days can legitimately be transmuted into five aeons of a billion years each? Any such conversion will require unique techniques of interpretation, designed expressly for this one chapter. Liberals are quick to ridicule those who accept what is called the "literal interpretation" of Genesis, forgetting that there is really no such thing! If something is "interpreted," it is not taken literally at all. An "interpretation" is actually a "translation," in which words are not taken at face value, but are converted into other words. Thus, "day" may be transformed into "age" and "all" into "some." If God actually employed such a coding technique in writing Genesis One, it is strange that He would withhold the key to the code from all His holy prophets and apostles until it was forced from Him, as it were, by the unbelieving scientists of the latter days.

When men use figures of speech in their writing,

they normally do it for emphasis and clarification, not for cryptic concealment. The creation account is clear, definite, sequential and matter-of-fact, giving every appearance of straightforward historical narrative. If it is an allegory, poem, liturgy, saga, or some other literary form, it is, to say the least, highly deceptive. If the Author did actually intend to tell of the sequential creation of all things in six *days,* it is impossible to see how He could have done so more plainly than in the words and phrases actually used. Would God be party to such a blatant deception? Perhaps it is not so "naive" after all to take Genesis "literally." Christ did.

But suppose we do take it symbolically anyhow, and let "day" equal "aeon." We may thereby solve the problem of duration but we still have the problem of sequence. That is, there are at least 25 discrepancies between the sequence of events in Genesis One and the evolutionary sequences of geology. For example, Genesis informs us that the sun, moon, and stars were made only during the fourth "aeon," half-way through geologic time, long after plant life in its highest forms evolved. Although theories of solar and stellar evolution have been numerous and varied, none has suggested such an idea as this. To resolve this discrepancy, we must again become interpretively ingenious. Perhaps the sun was not really made on the fourth day; rather the clouds in the sky were dispersed so that it could be seen! But, then, what was it that happened on the first day, when God said "Let there be light"?

Well, no doubt, only *some* of the clouds were dispersed then, the rest of them on the fourth day. God called this first light "good" and it was good enough to evolve plant life and maintain it for millions of years, but it wasn't yet good enough to satisfy God, so He finally decided to break up the rest of the cloud cover and let the sun, moon, and stars be seen on earth.

Of course, there was no man there to see them, nor would there be for many millions of years to come. The heavenly bodies were also to be used for "signs, and for seasons, for days and for years," but they wouldn't be useful for this purpose for some while yet. Maybe there was no need for such haste in breaking up that cloud cover after all!

Perhaps the above fantasy is not sufficiently respectful to the widely-held "day-age" theory, but this is only one of its scientific discrepancies, and there are two dozen more. The problem of the geologic ages does need to be resolved, but exegesis like this is *not* the solution.

What about placing the geologic ages somewhere *before* the six days of creation? Then we could accept the six days of creation literally, as the writer intended them. If we do this, however, we would have to assume that the plants and animals were only *re-created* at that time, since they had all previously existed (as shown by the fossils which identify these supposed pre-Adamic geologic ages) and presumably been destroyed in a mysterious worldwide cataclysm which left the earth devastated and in gross darkness, as described in Genesis 1:2.

We still have this troublesome business about having to disperse some of the clouds on Day One, and the rest on Day Four, of the re-creation week. And what about the "waters above the firmament"? Were they part of these clouds? They were installed on Day Two. Perhaps this proved to be a mistake, since they reversed part of the cloud dispersal on Day One. What purpose they may have served, if any, is unrevealed, since they, along with the rest of the primeval cloud bank, were all apparently dissipated two days later. They could hardly have been in the vapor state, since the atmosphere was already saturated when they were elevated.

But the absolute darkness of Day One could not have been caused by water clouds anyhow. Perhaps the clouds were clouds of dust and debris, hurled into the atmosphere by a gigantic global explosion which disintegrated the earth's crust, boiled away its oceans and destroyed all life on its surface. No lesser cataclysm could conceivably produce such thick darkness, nor require such extensive repair work from God as described in the six days of re-creation.

The main reason for all this speculation, of course, is to try to make room for the geologic ages before Genesis. The geologic ages, however, are synonymous with the fossil record in the sedimentary layered rocks of the geologic column, the great thicknesses of hardened and stratified sediments which now comprise the earth's crust. If we destroy the sediments and the fossils by our hypothetical cataclysm (and the type of cataclysm required would undoubtedly do that) then we don't have any geological ages left anyhow!

But suppose that, by some miracle, the fossils and sediments were all left intact by the cataclysm. Then, whence came all the debris with which to blot out the sun and leave the earth in pitch blackness? And, furthermore, where is the geologic evidence for such a cataclysm? Anything of such global extent and such devastating effects surely must have left a clear record of some kind in the earth's crust. But orthodox geology recognizes no worldwide cataclysms, least of all one in the most recent of geologic times.

Some might suggest that the entire fossil-bearing geologic column was itself formed by the cataclysm (indeed we believe this may actually have been the case in the great Flood described in Genesis 6–9), but if that were the case then there remains no evidence for the geological ages. It should be emphasized again that it is the fossil-bearing sedimentary rocks of the geologic column that provide the evidence for the geologic ages. If all the fossils were buried in one cataclysm, there remains no evidence for any pre-Adamic ages, and therefore no need for the cataclysm! The theory is self-negating.

Thus neither the day-age theory nor the pre-Adamic cataclysm theory is capable of harmonizing the Genesis record with the scientific implications of the geologic-age concept. Not only are these two ideas unscientific, however—they are also un-Scriptural. For example, Exodus 20:8–11 reads: "Remember the sabbath day, to keep it holy. Six days shalt thou labor, and do all thy work: But the seventh day is the sabbath of the Lord thy God: . . . For in six days the Lord made heaven and earth, the sea, and all that in them is, and rested the seventh day: wherefore the Lord blessed the sabbath day, and hallowed it."

These words were written on a tablet of stone by the finger of God Himself (Exodus 31:18) and thus certainly must be taken with all seriousness. The basis of the fourth commandment, dealing with the six days of man's work week, is that God did all *His* work in six days. The "days" are obviously the same kind of "days" in both cases, or else God is incapable of meaningful communication. The word "days," in the plural like this, (Hebrew *yamim*) occurs approximately 700 times in the Old Testament, and *always* means literal days.

Furthermore, not only were the "days" real days, but the work of the six days was all-embracing. "Heaven and earth, the sea, and all that in them is," were made in the six days. There could therefore have been no part of any pre-existing earth left over. "All that in them is" presumably means "all that in them is," since God Himself said so. That being the case, there could be no left-over fossils, rocks, radioactive minerals or anything else that could give any indication of a previous era.

There are numerous other exegetical problems with both the day-age and pre-world hypotheses, but the most serious fallacy in both of them is that they charge God with stupidity and cruelty. Since the very *raison d'etre* for either of these devices is to retain the geologic age system, they necessarily incorporate the fossil record, with its apparent billion-year reign of random change, struggle, disease, suffering and death, all over the world. The appalling inefficiency and barbarity of the evolutionary process is evidence enough that God would not allow, let alone invent, such a thing.

The sufferings of this present world, of course, are because of sin and the Curse. The geologic ages, however, with all their suffering and death, presumably took place long before man sinned. Nor does it help to suggest that Satan's sin may somehow be responsible. His sin did not take place until *after* the geologic ages, not before. The pre-Adamic cataclysm, in fact, is explained by its advocates as the direct and immediate result of Satan's sin. Thus the sin of Satan in no way accounts for the suffering and death of the geologic ages which preceded it. As a matter of fact, it is inconceivable that God, at the end of the six days, could have pronounced "everything that He had made" to be "very good" (Genesis 1:31) if, at the time, the earth was scarred with ages of violence and death, and the heaven itself was a battleground between the devil and his angels and Michael and his angels. As far as any clear evidence from Scripture is concerned, the sin and fall of Satan did not occur until after man's creation, chronologically somewhere between the end of Genesis 2 and the beginning of Genesis 3.

Scientifically, exegetically and theologically, therefore, it is impossible to retain the geological-age system along with the six days of creation. The Bible *does* teach, after all, that all things were created in six real days several thousand years ago. Maybe this is "naive literalism," but it is what God has said. It is not merely the oft-ridiculed "literal interpretation" of Genesis; it is the actual meaning of the words themselves and no other interpretation fits the context.

A good question is: if it really took five billion years for God to make all things, why did He tell us it took six days?

We are forced to the conclusion, as Bible-believing Christians, that the earth is really quite young after all, regardless of the contrary views of evolutionary geologists. This means then that all the uranium-lead measurements, the potassium-argon measurements and all similar measurements which have shown greater ages have somehow been misinterpreted.

It should be remembered that history, in the sense of written records, supports the Bible chronology. The first dynasty of Egypt, the first kings of Sumeria, and all other confirmed dates in the history of man, indicate that civilization began in the Middle East, somewhere near Mount Ararat, about four or five thousand years ago. All earlier dates have been estimated indirectly by some physical process—radiocarbon decay, uranium decay, or something else. Any such indirect process necessarily must be based on certain assumptions which, in the nature of the case, could never be proved to be true.

How, then, can they be reinterpreted? Is it possible that other assumptions could be made which, with the same physical data, will give ages that accord with the Biblical chronology?

It is true, of course, that even evolutionary scientists recognize that many problems and errors can affect these measurements. The techniques themselves are very difficult and are easily subject to large experimental error. The systems that are used are readily contaminated or influenced by external factors (e.g., uranium may be leached out of a uranium-lead system, free argon may be trapped into a potassium-argon system, etc.). The process rate itself may have varied from its present rate at some time or times in the past. A considerable number of volcanic rocks, known to have been formed by volcanic lava flows in modern times, have been "dated" by potassium or uranium techniques to be millions or billions of years in age.

Many, perhaps most, geologic dates measured by such processes are subject to considerable uncertainty because of problems such as these. But it does still seem that so many of them give dates that are far greater than the Bible would allow that there must be some other more basic reason for the discrepancy.

It is noteworthy that only a relatively small number of physical processes have been adopted for use as geologic "clocks." In principle, there should be almost an infinite number of processes available for such use. Every process involves some kind of change with time and therefore could theoretically be used to measure the passage of time. Why, then, have only those few been selected that change very slowly with time?

There are numerous processes which, on the basis of the same kinds of assumptions that are commonly used for uranium dating and similar dating methods, would give vastly smaller periods of time—thousands or millions of years, rather than billions. Why are these never used? Most people, even most scientists, apparently have never even considered this question. Somehow the natural man reacts favorably to any evidence that the earth is very old, but unfavorably to any indication that it may be young.

If the earth is really only several thousand years old, as the Bible teaches, then there obviously is no time for any significant evolutionary process to have occurred. Creation is the only alternative, and men are very uncomfortable when made to realize God has a direct and immediate concern with this world, with the development of human events, and with their own personal lives. Consequently, they seek by every means available, consciously or subconsciously, to relegate God to as inconspicuous a role as possible—as far removed in time and space and interest from the physical universe as the data will possibly allow.

This purpose is nicely served by the scientific premise of uniformitarianism, which holds that the origin and development of all things can be described in terms of the same natural laws and processes which function today. Thus there is no room for either a period of special creative processes or a period of special catastrophic processes in the past. If the world must have attained its present state by the same slow processes which operate today, then an immensity of time is demanded.

Even on the basis of uniformitarian assumptions, it would seem that the processes which yield young ages are more likely to be correct than those which give great ages. That is, a process is more likely to operate uniformly and without external interruptions for a short period of time than for a long period of time. Nevertheless, the only methods that have been acceptable to uniformitarians have been those which give long ages.

Other processes do give much younger ages, however. For example, the present rate of sedimentary erosion would have reduced the continents to sea level in six million years and would have accumulated the entire mass of ocean-bottom sediments in 25 million years. Present rates of volcanic emissions would have produced all the water of the oceans in 340 million years and the entire crust of the earth in 45 million years.

There is no measurable accumulation of meteoritic dust on the earth's surface, but present rates of influx of such dust from space would produce a layer ⅛ inch thick all over the earth in a million years and a layer 54 feet thick in 5 billion years. The comets of the solar system are disintegrating so rapidly that they could only have come into existence less than a few million years ago at most. The earth's magnetic field is decaying so rapidly that its origin cannot have been more than about 10,000 years ago.

The rate of uranium influx into the ocean indicates a maximum age for the oceans of about one million years. Sodium influx indicates perhaps 100 million years, but chlorine, sulfates and other materials give much less. Helium influx to the atmosphere shows the maximum age of the atmosphere to be about 12,000 years.

As far as human history is concerned, trends of population statistics show the most probable date for the beginning of human populations to be about 4,000 years ago. Radiocarbon dating, if rightly understood in terms of a non-equilibrium dating equation (instead of the steady-state model which is commonly used despite strong evidence that about 25% more radiocarbon is being formed in the atmosphere than is decaying in the biosphere) will show that the maximum age of the atmosphere and of any archaeological site is about 7,000 years.

The above calculations are all based on uniformitarian assumptions and, therefore, in most cases give ages that are too large. However, they are all at least as accurate and reliable as the various uranium and potassium dates that suggest an age of several billion years for the earth.

Aside from this question of arbitrary screening of dates and dating methods, however, how is it that some processes do at least fairly consistently give such great ages?

The answer probably lies in a fuller comprehension of the effects of Creation on the earth's structure and the Flood on the earth's processes. The Apostle Peter suggests this in his great farewell chapter, II Peter 3. He first predicted that the present attitude of intellectuals in the last days would be that of evolutionary naturalism, as expressed succinctly in their dogma of uniformitarianism: "All things continue as they were from the beginning of the creation."

He then charges that his philosophy is based on a deliberate rejection of the fact of a primeval special creation and a later world-destroying deluge (II Peter 3:5,6). The fact of an initial complete creation means that all the earth's components (including mineral systems) were created in a fully-developed, balanced, harmonious complex, perfectly integrated and distributed through the earth in accordance with God's own purposes. The fact of a subsequent global cataclysm means that the continuity of all natural processes was drastically interrupted and, in many cases, the process rates vastly accelerated during the Flood epoch.

Wherever radioactive minerals were placed in the primeval earth, therefore, they were probably also placed in association with their "daughter" elements in equilibrium and concordant amounts. Their "apparent age" at the completion of creation was a function of the processes of their creation, not of decay processes inaugurated *after* the creation. There is no way we can examine these creative processes scientifically, since "God rested," and such processes no longer exist.

The initial "apparent age" of all such systems, has, of course, been affected by the earth's subsequent history, especially the Curse and the Flood. During and after the Flood, there were great magmatic flows, great earth movements, great volcanic eruptions, great atmospheric disturbances, and other phenomena which catastrophically changed all previous relationships. Since most uranium, rubidium, and potassium minerals as used for dating are found in igneous intrusives, volcanic basalts, or transported sedimentary rocks, it is no wonder that most such minerals give such an array of discordant and anomalous ages as they do. The few that give concordant results may represent exceptional instances of undisturbed transportation and redisposition during the cataclysmic upheavals of the Flood epoch.

In any case, it is evident that application of a uniformitarian calculation to such systems can yield no information whatever about their *true age,* even though it may be possible occasionally to get an approximate indication of their initially created "apparent age." The only way we can determine the true age of the earth is for God to tell us what it is. And since He *has* told us, very plainly, in the Holy Scriptures that it is several thousand years in age, and no more, that ought to settle all basic questions of terrestrial chronology.

The Apostle Peter has another word in this connection: "Beloved, be not ignorant of this one thing, that one day is with the Lord as a thousand years" (II Peter 3:8). In context, he is saying: "Remember that God is not limited in the accomplishment of His work by uniformitarian rates, as naturalistic scoffers have assumed. God can do in one day what might seem to men to require a thousand years." Therefore do not be impressed by the "apparent age" of prehistoric formations. The "true age" is what God says it is, and there is no other way of determining it.

Then he interjects a sober warning and admonition: "They that are unlearned and unstable (that is, in the doctrinal understanding of God's Word) wrest . . . the Scriptures, unto their own destruction. Ye, therefore, beloved, seeing ye know these things before, beware, lest ye also, being led away with the error of the wicked, fall from your own steadfastness" (II Peter 3:16,17).

Instead of such a dreary strategy of continual retreat and reinterpretation and apostasy, he concludes: "But grow in grace, and in the knowledge of our Lord and Saviour Jesus Christ. To Him be glory both now and forever. Amen" (II Peter 3:18).

APPENDICES

APPENDIX A-1
BIBLICAL ARGUMENTS FOR A WORLDWIDE FLOOD

Genesis Text	Argument
1. 1:7	Water above the atmosphere must have been global in extent.
2. 2:5	No rain upon the earth must have been worldwide condition.
3. 2:6	Earth mist watered the whole face of the ground.
4. 2:10-14	Edenic geography no longer in existence.
5. 4:22	High civilization at dawn of history not continuous with present world.
6. 5:5, etc.	Longevity of antediluvian patriarchs indicates distinctive biosphere.
7. 6:1	Man had multiplied on the face of the earth.
8. 6:2	Demonic-human unions coextensive with mankind.
9. 6:5	Universal evil inexplicable in post-diluvian society.
10. 6:6, 7	Repentance of God extended to the whole animal creation.
11. 6:11	Earth was filled with violence and corruption before God.
12. 6:12	All flesh was corrupted (possibly including animals).
13. 6:13	God decided to destroy both man and the earth.
14. 6:15	Ark too large for regional fauna.
15. 6:17	Everything with the breath of life to die.
16. 6:19	Purpose of ark was to keep two of every sort alive.
17. 6:20	Animals of all kinds migrated to the ark.
18. 6:21	All kinds of edible food taken on the ark.
19. 7:4	Every living substance on the ground to be destroyed.
20. 7:10	"The flood" (Hebrew *mabbul*) applies solely to Noah's flood.

21.	7:11	All the fountains of the great deep cleaved open in one day.
22.	7:11	The "sluiceways from the floodgates" of heaven were opened.
23.	7:12	Rain poured continuously for forty days and forty nights.
24.	7:18	The waters prevailed and increased greatly.
25.	7:19	High hills under the whole heaven were covered.
26.	7:20	Waters fifteen cubits above highest mountains.
27.	7:21	Every man died on the earth.
28.	7:22	All flesh with the breath of life in the dry land died.
29.	7:23	Every living substance destroyed off the face of the ground.
30.	7:24	Waters at maximum height for five months.
31.	8:2	Fountains of deep open for five months.
32.	8:2	Windows of heaven open for five months.
33.	8:4	Ark floated over 17,000 ft. mountains for five months.
34.	8:5	Water receded 2½ months before mountain tops seen.
35.	8:9	Dove found no suitable ground even after four months of recession.
36.	8:11	Plants began budding after nine months of the flood.
37.	8:14	Occupants were in the ark over a year.
38.	8:19	All kinds of present non-marine animals came from the ark.
39.	8:21	God smote all things living only once.
40.	8:22	Present uniformity of nature dates from the end of the flood.
41.	9:1	Earth was to be filled with descendants of Noah.
42.	9:2	Changed relation between man and animals followed the flood.
43.	9:3	Man permitted animal food after flood.
44.	9:6	Institution of human government dates from flood.
45.	9:10	God's covenant made with every living creature.
46.	9:11	The flood promised by God never to come again on the earth.
47.	9:13	Rainbow placed in sky after the flood.
48.	9:19	Whole earth overspread of the sons of Noah.
49.	11:1	Whole earth of one language after the flood.
50.	11:9	All men lived in one place after the flood.

Other Texts

1.	Job 12:15	The waters overturned the earth.
2.	Psalm 29:10	The flood testified God as eternal king.
3.	Psalm 104:8	Flood terminated by crustal tectonics.
4.	Isaiah 55:9	Waters of Noah went over the earth.
5.	Matthew 24:37	The days of Noah like those when Christ comes.

6.	Matthew 24:39	The flood took them all away.
7.	Luke 17:27	The flood destroyed them all.
8.	Hebrews 11:7	Noah warned of things never seen before.
9.	Hebrews 11:7	Noah condemned the world by his faith.
10.	I Peter 3:20	Only eight souls saved on the ark through the flood.
11.	II Peter 2:5	God spared not the old world (Greek *kosmos*).
12.	II Peter 2:5	God brought the flood on the world of the ungodly.
13.	II Peter 2:5	The "flood" (Greek *kataklusmos*) applied solely to Noah's flood.
14.	II Peter 3:6	The world that then was perished by the watery cataclysm.

APPENDIX A-2
NON-BIBLICAL ARGUMENTS FOR WORLDWIDE FLOOD

1. Worldwide distribution of flood traditions.
2. Origin of civilization near Ararat-Babylon region in post-flood time.
3. Convergence of population growth statistics on date of flood.
4. Dating of oldest living things at post-flood time.
5. Worldwide occurrence of water-laid sediments and sedimentary rocks.
6. Recent uplift of major mountain ranges.
7. Marine fossils on crests of mountains.
8. Evidence of former worldwide warm climate.
9. Necessity of catastrophic burial and rapid lithification of fossil deposits.
10. Recent origin of many datable geological processes.
11. Worldwide distribution of all types of fossils.
12. Uniform physical appearance of rocks from different "ages".
13. Frequent mixing of fossils from different "ages".
14. Near-random deposition of formational sequences.
15. Equivalence of total organic material in present world and fossil world.
16. Wide distribution of recent volcanic rocks.
17. Evidence of recent water bodies in present desert areas.
18. Worldwide occurrence of raised shore lines and river terraces.
19. Evidence of recent drastic rise in sea level.
20. Universal occurrence of rivers in valleys too large for the present stream.

21. Sudden extinction of dinosaurs and other prehistoric animals.
22. Rapid onset of glacial period.
23. Existence of polystrate fossils.
24. Preservation of tracks and other ephemeral markings throughout geologic column.
25. Worldwide occurrence of sedimentary fossil "graveyards" in rocks of all "ages".
26. Absence of any physical evidence of chronologic boundary between rocks of successive "ages".
27. Occurrence of all rock types (shale, limestone, granite, etc.) in all "ages".
28. Parallel of supposed evolutionary sequence through different "ages" with modern ecological zonation in the one present age.
29. Lack of correlation of most radiometric "ages" with assumed paleontologic "ages".
30. Absence of meteorites in geologic column.
31. Absence of hail imprints in geologic column, despite abundance of fossil ripple-marks and raindrop imprints.
32. Evidence of man's existence during earliest of geologic "ages" (e.g., human footprints in Cambrian, Carboniferous, and Cretaceous formations).

APPENDIX B
NEW TESTAMENT REFERENCES TO GENESIS 1-11

New Testament Reference	Topic	Genesis Reference
1. Matthew 19:4	Created male and female	5:2
2. Matthew 19:5, 6	Cleave unto his wife	2:24
3. Matthew 23:35	Righteous Abel	4:4
4. Matthew 24:37-39	Days of Noah	6:3-5
5. Mark 10:6	God made them	1:26, 27
6. Mark 10:7-9	One flesh	2:24
7. Mark 13:19	Creation which God created	2:4
8. Luke 1:70	Prophets since the world began	4:26
9. Luke 3:34-36	Son of Thara . . . Son of Sem	11:10-24
10. Luke 3:36-38	Son of Noe . . . Son of Adam	5:3-29
11. Luke 11:50-51	Blood of Abel	4:8-11
12. Luke 17:26-27	The Flood came and destroyed them all	7:10-23

13. John 1:1-3	In the beginning God	1:1
14. John 1:10	World made by Him	2:3
15. John 8:44	Father of lies	3:4, 5
16. Acts 3:21	Restoration of all things	5:29
17. Acts 4:25	All that in them is	2:1
18. Acts 14:15	Fruitful seasons	8:21, 22
19. Acts 17:24	God made all things	1:31
20. Acts 17:26	All nations on face of the earth	10:32
21. Romans 1:20	Things that were made	2:4-6
22. Romans 5:12	Death by sin	2:17
23. Romans 5:14-19	Death reigned from Adam	4:5-31
24. Romans 8:20-22	Bondage of corruption	3:17-18
25. Romans 16:20	Satan bruised under foot	3:15
26. I Corinthians 6:16	One flesh	2:24
27. I Corinthians 11:3	Head of the woman	3:16
28. I Corinthians 11:7	In the image of God	1:27
29. I Corinthians 11:8, 9	Woman of the man	2:23
30. I Corinthians 15:21, 22	By man came death	3:19
31. I Corinthians 15:38, 39	Every seed his own body	1:11, 21, 24
32. I Corinthians 15:45	Adam a living soul	2:7
33. I Corinthians 15:47	Man of the earth	3:23
34. II Corinthians 4:6	Light out of darkness	1:3-5
35. II Corinthians 11:3	Serpent beguiled Eve through subtlety	3:1-6
36. Galatians 4:4	His Son, made of a woman	3:15
37. Galatians 4:26	Mother of us all	3:20
38. Ephesians 3:9	Created all things	2:3
39. Ephesians 5:30, 31	Bone of His bone	2:23, 24
40. Colossians 1:16	All things created	2:1-3
41. Colossians 3:10	Created in His image	1:27
42. I Timothy 2:13-15	Bone of His bone	2:23, 24
43. Hebrews 1:10	Earth and heavens in the beginning	1:1
44. Hebrews 2:7, 8	All things in subjection under Him	9:2
45. Hebrews 4:3	Words were finished	2:1
46. Hebrews 4:4	Rest on the seventh day	2:2
47. Hebrews 4:10	Ceased from His works	2:3
48. Hebrews 11:4	Abel a more excellent sacrifice	4:3-5
49. Hebrews 11:5	Enoch translated	5:21-24
50. Hebrews 11:7	Noah saved his house	7:1
51. Hebrews 12:24	Blood of Abel	4:10
52. James 3:9	Men in the similitude of God	5:1
53. I Peter 3:20	Preparing of the ark	6:14-16
54. II Peter 2:4	Angels that sinned	6:1
55. II Peter 2:5	God spared not the old world	6:8-12
56. II Peter 3:4, 5	Out of the water and in the water	1:5-7

57.	II Peter 3:6	Overflowed with water	7:17-24
58.	I John 3:8	Devil sinneth from the beginning	3:14
59.	I John 3:12	Cain slew his brother	4:8, 25
60.	Jude 6	Angels left their own habitation	6:4
61.	Jude 11	The way of Cain	4:16
62.	Jude 14, 15	Enoch, the seventh from Adam	6:18-24
63.	Revelation 2:7	Tree of life	2:9
64.	Revelation 3:14	Beginning of the creation of God	3:14
65.	Revelation 4:11	Created all things	2:3
66.	Revelation 10:6	Created heaven and earth and all things	2:1
67.	Revelation 12:1-4	Seed of the woman	3:15
68.	Revelation 12:9	That old serpent	3:14
69.	Revelation 12:13-17	Enmity between thee and the woman	3:15
70.	Revelation 14:7	He that made heaven and earth	2:4
71.	Revelation 17:5	Babylon, the mother of abominations	10:8-11
72.	Revelation 17:18	That great city	11:4, 5
73.	Revelation 20:2	The dragon, that old serpent	3:1
74.	Revelation 21:1	First heaven and first earth	2:1
75.	Revelation 21:4	No more death, or sorrow, or crying, or pain	3:17-19
76.	Revelation 22:2	Fruit of the tree of life	3:22
77.	Revelation 22:3	No more curse	3:14-19

NOTE: All New Testament books except Philippians, I and II Thessalonians, II Timothy, Titus, Philemon, II and III John have references to Genesis 1-11. Every chapter of Genesis 1-11 is referred to somewhere in the New Testament. Every New Testament writer refers to Genesis 1-11. Jesus Christ referred to Genesis 1, 2, 3, 4, 5, 6, and 7.

INDEX OF SCRIPTURES

INDEX OF SUBJECTS

RECOMMENDED BOOKS FOR FURTHER READING

Available From

Creation-Life Publishers, Inc.
P. O. Box 15666
San Diego, California 92115

Other Books by Henry M. Morris, Ph.D.
(Director, Institute for Creation Research)

SCIENTIFIC CREATIONISM
Most comprehensive, documented exposition of all the scientific evidence of origins. Scientific data presented without bias, followed (in the general edition only) by an extensive discussion of the biblical aspects of creationism.

General Edition (biblical documentation) No. 140
Public School Edition (non-religious text) No. 141

THE GENESIS FLOOD
(Coauthor John C. Whitcomb, Jr., Th.D.)
The standard classic text in the field of scientific biblical creationism and catastrophism. Thoroughly documented treatment of the biblical and scientific implications of creation and the flood. Widely recognized as the most authoritative book on this subject. No. 069

THE TROUBLED WATERS OF EVOLUTION
In addition to presenting a nontechnical study of the evidence for creation, this book traces the history of evolutionary thought and specifies the areas of our society that have been devastated by it. No. 170

MANY INFALLIBLE PROOFS

One of the best in its field. As it scripturally supports each basic tenet of the Christian faith, it strengthens new Christians, as well as convincing the unbeliever of the integrity of God's Word. This is a real help to personal growth and effective witnessing in today's skeptical society. This volume has been adopted as a complete textbook for college courses in apologetics and Christian evidences. Kivar, No. 102; Cloth, No. 103

THE GENESIS RECORD

A one-of-a-kind scientific and devotional commentary on the book of beginnings. Both theologians and lay persons will gain great understanding of this foundational book of the Bible. Cloth, No. 070

KING OF CREATION

This unique book places the modern creation movement in its biblical perspective, emphasizing Christ as Creator and Sovereign of the world. It documents the scientific strength and spiritual impact of creation, while reaffirming the necessity for all Christians to become actively involved in preaching the "gospel of creation."

No. 096

THE BIBLE HAS THE ANSWER

(Coauthor Martin Clark, D.Ed.)

Scientific, logical, and biblical answers to 150 of the most frequently asked questions on the Bible and science, evolution, supposed mistakes in Scripture, difficult doctrines, social problems, and practical Christian living. Complete topical and Scripture indexes. On doctrinal questions, orientation is pre-millennial and Baptistic; on others, non-denominational. No. 023

THAT YOU MIGHT BELIEVE
This book has been instrumental in winning countless souls to Christ. It contains numerous examples of archaeological and scientific documentation of many incidents recorded in the Bible—special creation, the Flood, fulfilled prophecy, and the eternal plan of God. If proof of the authenticity of the Bible is necessary to confirm God's existence and integrity, *here is that proof.* No. 168

THE BIBLE AND MODERN SCIENCE
Evangelistic presentation of evidences for the scientific validity of the Bible, including historical and prophetic confirmation. No. 333

EVOLUTION AND THE MODERN CHRISTIAN
Popular-level exposition of evidence for creation versus evolution. Especially written for young people. No. 344

STUDIES IN THE BIBLE AND SCIENCE
Sixteen studies on special topics, including the Bible as as scientific textbook, evidence of Christ and the Trinity in nature, biblical hydrology, concept of power in Scripture, scientism in historical geology, and others. No. 377

THE SCIENTIFIC CASE FOR CREATION
This is a brief introduction to the broad field of Scientific Creationism. It is treated solely from a scientific point of view, containing no theological implications. Although some scientific background on the part of the reader is helpful, an attempt has been made to enable the general reader to follow the arguments and appreciate the conclusions. No. 139

Books by Duane T. Gish, Ph.D.

(Associate Director, Institute for Creation Research)

EVOLUTION? THE FOSSILS SAY NO!

The only *solid* evidence in the discussion of origins is the fossil record—anything else is circumstantial evidence and conjecture. Powerful testimony about the origin of our earth's inhabitants.

General Edition No. 054

Public School Edition (non-religious text) No. 055

DINOSAURS: THOSE TERRIBLE LIZARDS

Did dinosaurs and humans live at the same time? Are dragons just imaginary? Does the Bible describe dinosaurs? This beautifully color-illustrated book for children tells about dinosaurs and why they no longer exist. 8½" x 11". Cloth, No. 046

MANIPULATING LIFE: WHERE DOES IT STOP?

(Coauthor Clifford Wilson, Ph.D.)

Two top scientists and educators discuss the ethical and moral aspects of biological engineering, cloning, test-tube babies, surrogate motherhood, abortion, recombinant DNA, and genetic manipulation. Can man "control his own evolution?" Can he create supermen —or superclones . . . or will he ultimately create monsters? Cloth, No. 100

Other Books

HE WHO THINKS HAS TO BELIEVE

A. E. Wilder-Smith, Ph.D., Dr. es Sc., D.Sc., F.R.I.C.

Explores the marvelous process of reasoning and drawing logical conclusions, with the result that anyone who uses these natural abilities of the mind properly *must* arrive at the conclusion that there is a Creator-God.

No. 077

WHAT'S IN AN EGG?
Joan Gleason Budai
Many things get their start in eggs. This book lays the groundwork for a truthful, open discussion when asked, "Where did I come from?" Through pictures and plain talk, *What's In An Egg?* tells the story from "egg" to actual birth about fowl, mammals, insects, fish, and many other of God's marvelous creations.
No. 185

WHY DOES GOD ALLOW IT?
A. E. Wilder-Smith, Ph.D., Dr. es Sc., D.Sc., F.R.I.C
If there is a God—who is supposed to be "good" and loving—why does He permit all the violence and suffering in the world? Does the abundance of evil prove He really doesn't exist? Sensible answers to questions that have plagued people since the beginning. Gripping photos.
No. 186

DRY BONES . . .And Other Fossils
Gary E. Parker, M.S., Ed.D.
What are fossils? How are they formed? What can we learn from them? Answers in conversational dialogue in this creatively illustrated book for children. 8½" x 11".
No. 047

TRACKING THOSE INCREDIBLE DINOSAURS
(And The People Who Knew Them)
John D. Morris, Ph.D.
What's the *real* story on those footprints in the Paluxy River bed? Are they authentic? What do they really tell us? A first-hand report, documented by nearly 200 photographs.
No. 173

CREATION: THE FACTS OF LIFE
Gary E. Parker, M.S., Ed.D.
If you are an "armchair scientist" with a hunger for knowledge (but no Ph.D. to help you understand it!), this book was written for you. Dr. Parker has clearly explained the basic concepts of creation and evolution, including "down home" examples to illustrate many ideas that, until now, have been beyond the grasp of many readers. "Science for the *un-scientist.*" No. 038

APE-MEN: FACT OR FALLACY?
Malcolm Bowden
Remember those colorful drawings supposedly tracing man's ancestors from the monkey stage to the man stage? Well, the myth has been destroyed once and for all as Mr. Bowden presents detailed evidence that these and other discoveries of "ape men" were, in fact, imaginative reconstructions and forgeries. Fossils provide the facts concerning the ancestry of man; there is no longer any room for fantasy. Ape men are real only in the imaginations of those who wish to avoid the facts.
No. 006

AN ANTHOLOGY OF A DECADE OF CREATIONIST ACTIVITY
The past decade has seen a great increase in interest and controversy in the realm of creationism. These four books trace this growth through the compilation of significant articles and debates that have been reported in ICR's popular *Acts & Facts* publication. Although they may be ordered individually, the complete 4-volume set provides a comprehensive record of the recent revival of interest in creation science, as well as a complete set of the valuable "Impact Articles".

Creation: Acts/Facts/Impacts **No. 037**
The Battle for Creation **No. 013**
Up With Creation **No. 179**
Decade of Creation **No. 044**

EVOLUTION: ITS COLLAPSE IN VIEW?
Henry Hiebert
In brief but comprehensive statements, the author discredits all of the claims of evolution, telling *what* is false and *why* it is false. This capsulized cross-examination of the tenets of evolution is a very readable and understandable tool for laymen, as well as high school and college students. No. 057

THE MOON: ITS CREATION, FORM, AND SIGNIFICANCE
John D. Whitcomb, Th.D., and
Donald B. DeYoung, Ph.D.
What is the moon really made of? How did it originate? How does it control the ocean tides? This book is packed with information that documents and explains recent discoveries made on our nearest space neighbor—the moon. Numerous charts, illustrations, and impressive full-color photographs make this study of the moon's history, significance, and destiny extremely fascinating.
Cloth, No. 105

CREATION OF LIFE
A. E. Wilder-Smith, Ph.D., Dr. es Sc., D.Sc., F.R.I.C.
Evaluates the practicing evolutionist's experimental design and data. Points out that scientific materialism holds the key to *neither* man's origin *nor* his destiny. Emphasizes that in order to have an efficient *design,* you must first have an efficient *designer*. No. 039

THE WATERS ABOVE
Joseph C. Dillow, B.S., Th.D.
Delineates in detail the physics and meteorology of such a vapor canopy and its maintenance in the antediluvian atmosphere. Before the *science* of such a canopy is considered, the author first establishes that such a canopy existed as taught by the biblical writers.
Cloth, No. 182

SHOULD EVOLUTION BE TAUGHT?
John N. Moore, Ed.D.
Discussion of the general and special theories of evolution and their philosophic and scientific character, respectively. Demonstration that the scientific creation model is better supported by facts than the evolution model. No. 142

THE GREAT BRAIN ROBBERY
David C. C. Watson
Is evolution a fact or a theory? Have we been brainwashed on the subject, robbed of our ability to think through the issues? This entertaining and infectiously enthusiastic, yet scholarly, book is short with nontechnical material which busy people can grasp quickly . . . at a bus stop or on the train. No. 074

SCIENCE IN THE BIBLE
Jean Sloat Morton, Ph.D.
The Bible's *scientific* statements or allusions are examined in the light of contemporary scientific knowledge. Subjects include astronomy, meteorology, chemistry, oceanography, earth science, zoology, anatomy, medicine, and diseases. Each scientific fact is illustrated. 119 illustrations, 15 of which are in full color.
Cloth, No. 137

THE NATURAL SCIENCES KNOW NOTHING OF EVOLUTION
A. E. Wilder-Smith, Ph.D., Dr. es Sc., D.Sc., F.R.I.C.
Internationally renowned scientist examines the evidence and presents the conclusions in this comprehensive analysis of evolution from the standpoint of the Natural Sciences. Recommended for teachers and college students, as well as laymen with a special interest in the study of origins. No. 110

THE EARLY EARTH
John C. Whitcomb, Jr., Th.D.
Studies in special topics relating to the Genesis record of creation, including the origin and nature of man, critical analysis of the gap theory, and others. Illustrated.
No. 051